Reclaiming Tomorrow's Climate: One Small Step for Man

A Thought-Provoking Journey From Climate Despair To Action For All Generations And Genders Alike

SIMON LEE

Dedication

Monika Lee

10/19/1976 – 09/10/2022

My Wife, Monia

In Eternal Memory of Your Boundless Love, Enduring Inspiration, and Unwavering Motivation.

Author's Note

My upbringing in the countryside of rural Staffordshire, surrounded by dairy farms and sharecropping, alongside raising sheep, goats, and chickens, provided me with a rich and practical education in the importance of hard work, a deep love for nature, and the principle of sustainability. This connection to the land and its cycles, the understanding of using only what you need, and the importance of giving back to the earth are invaluable life lessons that many people seek to learn as they become more aware of environmental issues and the need for sustainable living practices.

Having a father who was ahead of his time in harnessing available resources and building renewable energy systems has been incredibly inspiring. He was, and remains, an early advocate for renewable energy and sustainable practices, demonstrating innovation and forward-thinking that is crucial today as we face global environmental challenges. His influence likely instilled in me a sense of responsibility and creativity in approaching sustainability and conservation.

This unique combination of traditional farming values and innovative approaches to renewable energy and sustainability places me in a solid position to contribute to current conversations about environmental stewardship, sustainable agriculture, and renewable energy solutions. Whether through personal projects, community involvement, or broader advocacy, my background equips me with the knowledge, experience, and perspective to make meaningful contributions to these critical discussions.

My upbringing has shaped my character and work ethic. It has given me a profound respect for the natural world and a commitment to living in a way that honors the principles of sustainability and conservation. I want this work to represent a powerful narrative that resonates with many people seeking to make positive changes in their lives and communities to address environmental challenges.

Preface

Time is Running Out

The world is changing—and the trajectory is alarming. Extreme weather, rising sea levels, and dwindling resources are no longer distant warnings—they are our reality. This book isn't just a wake-up call; it's a strategic guide to action and transformation.

Within these pages, you'll find the tools to navigate the complexities of climate change with precision and purpose. Discover where your efforts can have the most significant impact, and learn how to channel your energy effectively for maximum results. With expert insights and time-saving strategies, this book empowers you to make a meaningful difference.

Offering both guidance and inspiration, it guides you from understanding the science to implementing tangible changes for sustainable living. Don't let the flood of information overwhelm you—take control and become an agent of change.

This is more than just a book; it's a rallying cry. Join the fight for a sustainable future.

Contents

Introduction

Imagine a world where one season blurs into the edges of another, where the din of megalopolises drowns in nature's rage, and the wind brings not sweet nothings but only a warning. No, it is not fiction or a mere screenplay—our torturing reality comes closer with every new day. This is so much more than just another book on climate change; this is a passionate call to begin, for fundamental measures, an exciting journey through science, reasons, and causes that provoke the most burning problem in the world.

Climate change cannot be overstated; the high stakes were well summarized by Barack Obama, America's 44th President: "The world must come together to confront climate change. There is little scientific dispute that if we do nothing, we will face more drought, famine, and mass displacement that will fuel more conflict for decades."[1]

[1] Reuters. (2015). Obama: World must tackle climate change swiftly. Retrieved from https://www.reuters.com/article/us-obama-climatechange/obama-world-must-tackle-climate-change-swiftly-idUSKBN0LR1UB20150224

The consequences of inaction on climate change are rather dire when seen from a global perspective. Let us take into consideration a few possible impacts that may surface:

Droughts: Failure to act on climate change contributes to an escalation in the number of devastating droughts experienced worldwide. These are unreasonably extended periods of inadequate rainfall, significantly threatening agricultural products, water sources, and the ecological system.[2]

Famine: Lack of action on climate change will exacerbate food insecurity and, at times, cause famines when disaster strikes and concerted global efforts to curb climate change are not implemented on time. Any such warming is likely to affect rainfall patterns adversely and lead to increased variability in precipitation; this is coupled with the higher frequency and intensity of extreme weather events, disrupting crop production and causing food insecurity.[3]

Mass Displacement: Millions more will potentially be displaced if climate change continues with similar inaction and apathy. Rising sea levels, increased frequency of natural calamities, and loss of habitable land push people away from their homes, leading to serious social, economic, and political challenges.[4]

---

[2] IPCC. (2018). Global warming of 1.5°C. An IPCC Special Report on the impacts of global warming of 1.5°C above pre-industrial levels and related global greenhouse gas emission pathways, in the context of strengthening the global response to the threat of climate change. Retrieved from https://www.ipcc.ch/sr15/

[3] FAO. (2021). Climate change and food security: Risks and responses. Retrieved from http://www.fao.org/3/cb4246en/cb4246en.pdf

[4] UNHCR. (2019). Global trends: Forced displacement in 2018. Retrieved from https://www.unhcr.org/5d08d7ee7.pdf

Individuals may have concerns about the economic impacts of climate change mitigation measures. They may worry that transitioning to renewable energy sources or implementing sustainable practices will lead to job losses or hinder economic growth.[5]

Many people still need to learn about the severity and urgency of climate change. Limited access to information or misinformation can contribute to a lack of understanding and motivation to take action.[6]

However, practical local and individual mitigation actions need to be improved because, at the international level, there needs to be sufficient political will and potent policies against this worldwide changing reality. These include poor governmental support, weak regulations, and a lack of international cooperation.[7]

Addressing climate change can often feel overwhelming, but there is a practical solution that can guide individuals toward meaningful action. The book gives readers shortcuts and actionable insights on navigating the intricacies of climate change, thus having real effects and saving readers' time.

Readers get the benefits of dealing with the problem of climate change effectively by reading this book in the following ways:

---

[5] Oxfam. (2020). Confronting carbon inequality: Putting climate justice at the heart of the COVID-19 recovery. Retrieved from https://oxfamilibrary.openrepository.com/bitstream/handle/10546/621905/bp-confronting-carbon-inequality-080920-en.pdf

[6] Hart, P. S. (2021). Climate change communication: Challenges and opportunities. Oxford Research Encyclopedia of Communication. doi:10.1093/acrefore/9780190228613.013.1197

[7] IPCC. (2014). Climate change 2014: Synthesis report. Contribution of Working Groups I, II and III to the Fifth Assessment Report of the Intergovernmental Panel on Climate Change. Cambridge University Press.

Time-Saving Shortcuts: The book provides concise and accessible information, saving readers valuable time in understanding the key concepts and challenges of climate change.

Actionable Insights: Readers can take away practical strategies and methods to make a difference in their everyday lives. From energy-saving tips to sustainable lifestyle choices, the book offers many practices for readers to implement.

This book emphasizes that every effort should be streamlined for maximum impact. It helps to identify those areas in which readers can create the most difference and lets readers focus on things that fit their interests, skills, or resources.

Throughout the book, the reader is inspired to take informed action; precisely, by providing timely and correct information, the book empowers its readers with the knowledge to make informed decisions and meaningful engagements in the climate change initiative. It highlights the need to take the opportunity to make a difference urgently and underlines the relevance of their journey against climate change.

This book provides a practical solution to climate change—applicable in both senses: the need for practical action and the feeling most people have of having too much to do and too many things clamoring for their attention. Through practical action and relevant insight, these time-saving shortcuts will undoubtedly empower the reader. In acting upon these opportunities, readers can become part of the collective, making a difference in climate change.

CHAPTER 1
Unveiling The Climate Enigma

Any good mystery has to start with "whodunit?" But humanity's greatest challenge is "what," not "who." We can't deal with the problem if we don't even acknowledge that we have one.

This book plunges straight into the core matter: climate change. We usually take it too lightly. As former United Nations Secretary-General Kofi Annan stated in 2009, "We, the first generation to be aware of the threat of climate change, are also the last generation that can take action to address it." And the enormity of Annan's responsibility in the above statement is not lost. We are the first generation to know, informed through clear signs, that the climate is breaking down and the last to make a real difference.

Despite the inevitability, Annan laments that humankind can always diminish the gravity of threats: "We constantly underestimate the power of new technologies and overestimate the power of old institutions. We persistently underestimate the seriousness of threats and exaggerate our capacity to cope with them." His words highlight a fundamental fact—we still haven't realized this. We are lulled back into a sense of security held over from past times when this problem seemed more manageable.

This book is an eye-opener, reminding us not to underestimate the severity of climate change. In the forthcoming pages, we shall delve deeper into our challenges. Also, before that, we must face up to the elephant in the room—a possibility that has existed for far too long and does not show signs of coming to an end: the undeniable probability that we have downplayed this crisis. This book leads us to what we don't want to hear about—climate change being a considerable threat that calls for our immediate attention.

We need to shift the focus. It's not enough to highlight the dangers; we must also showcase the positive progress. This will inspire hope and action. But let's not shy away from the harsh reality—we face serious problems.

One solution is to target industries with high carbon footprints. If a suitable alternative does not exist, we can heavily tax them. Think of it like smoking—a financial disincentive to discourage a harmful habit. After all, smokers eventually pay for their healthcare costs. We may not be able to tax future environmental problems directly, but these taxes can fund solutions and prepare us for the impacts we can't avoid.

The urgency is undeniable. We have discussed climate change for decades, yet many perceive it as a distant threat. It is high time we woke up from that illusion. Erratic weather, intensifying storms, disastrous fires—all are wake-up calls. We look no longer to a catastrophic future vacant with almost no people alive; as the text rightly observes, climate change stakes its claim today. We are severely downplaying the gravity of the situation. We need a paradigm shift—a deeper understanding of the current threats alongside a focus on the positive changes we can make. Combining these elements can inspire action and build a more sustainable future.

The Urgency of Tackling Climate Change Now

Climate change is not a theoretical problem in a far-off place that belongs to our great-grandchildren. The escalating global weather patterns and the rising temperature of our planet indicate the severity of the issue today. More and stronger storms threaten all coastal communities; the impacts are apparent even inland. The significant consequences of climate change are not to be ignored for natural environments or human settlements.

Given the urgency of addressing climate change and global warming, it is critical to appreciate the science behind these facts. After all, climate science is carefully conducted research that has been pursued over the years with well-documented evidence. While changes in sunlight, volcanic eruptions, and seismic activity are natural factors that might have contributed to shaping our climate, the human factor, or human-induced contributing factor, has caused the most significant impact. The acts associated with climate change are the burning of fossil fuels and deforestation, the loss of forests, especially in places with abundant precipitation like the Amazon Rainforest, the loss of forests, the impulse provision of oxygen, and various other roles in carbon sequestration. The loss of forests with speed is disturbing the ecosystem and, eventually, the generation of oxygen vital to the planet.

Among them is the rise in flooding due to the melting of glaciers. Melting icebergs do not have direct consequences like volcanic eruptions, but their effects are far-reaching in small areas. The increase in sea levels is a significant danger to the most populated cities in almost every country since most people live near the coasts. The devastation caused by flooding extends beyond the destruction of infrastructure; it also disrupts agricultural areas surrounding cities, leading to crop failures and significant economic losses. The financial burden of such damage affects everyone, as insurance costs rise for all individuals, regardless of their proximity to the coast.

Additionally, the depletion of groundwater resources, illegal drilling, groundwater accumulations, and ignorance of natural water filters like the Everglades further compound the challenges posed by climate change. The draining of underground reservoirs, extensive interference with waterways, and the salted intrusion of freshwater bodies may sooner or later threaten to reduce the water supply and disrupt fragile ecosystems. The destruction of marshlands and the resultant loss of sea life and various vegetation worsen the imbalance in the natural environment.

Climate change is a serious and complex problem that must be addressed immediately. This is not an overstatement. It threatens coastal residential settlements and economies and creeps far beyond the coastline to human habitations across all landscapes.

The scientific evidence, based on years of research, emphasizes the urgency of action. This is more critical, so the honest answer is to take action on the very causes of climate change; only when combined with sustainable practices will it be possible to decrease its overall impact and protect fundamental resources for healthier lives.

Now, imagine what our world could become if we don't act. Think of this: what used to be a bustling coastal metropolis—Miami—now stands solitary, a ghost town. Beaches no longer exist, consumed by rising sea levels that have surged farther inland, tragically pushing people out of their coastal homes and forcing them to flee to higher or drier land. Solitary and empty—skyscrapers, deserted this time, only adding to the reminder of the ever-present, lost paradise.

Crop failures are a dire reality on a global scale. The intrusion of saltwater and staple crop failures destroys fertile farmland. Breadbasket areas' production cannot provide the necessary volumes for most to feed their populations, while food prices rise. Grocery shelves turn bare, replaced with price controls and despair.

That is until extreme weather events become the norm. The continuously rising temperatures torched their way across California, taking parts of America's wealthiest valleys and turning them into nothing short of dust bowls.

Wildfires rage unchecked, blanketing entire cities in smoke. In keeping with record-shattering hurricanes, a tattered Gulf Coast is reduced to rubble—a similar fate befalls the few remaining intact towns. What was once considered unthinkable quickly comes to pass.

Freshwater scarcity will intensify in the age of devastation ahead. Imagine that: the Colorado River shrinks and dries completely in the full sweep of its course; lakes become shallow pools that look like mere puddles; and rivers appear to be the merest trickles of their previous selves. Conflicts over it rip apart societies and throw our world into chaos.

This failure will undoubtedly come at a tremendous cost—counting the billions of lives lost. Given that there are millions and millions of people, hundreds of thousands of them will be displaced by climate change worldwide. With the overcrowding of camps, diseases will be severe, coupled with failing healthcare systems. Social instability and violence are fueled by widespread desperation.

It does not have to be something we acknowledge. With a firm belief that we can act, it can still be ultimately altered so that it does not lead to the dystopian nightmare you just read. Learn more about the steps people and societies can take to mitigate change while building sustainability that successively takes us on and about. Let us not leave our children in a world ravaged by our inaction.

***

Let us illustrate the domino effect of climate change with a real-world example. I remember diving on the Great Barrier Reef several years ago. It is a heartbreaking shadow of its former glory, and scientists predict it could vanish entirely within a decade.

Rising water temperatures are the culprit. A mere 2-degree increase disrupts the delicate balance of the reef ecosystem. Coral, the foundation of the reef, bleaches and dies. This devastates the entire food chain—the vibrant fish populations that rely on the coral for habitat and food disappear.

However, the impact goes far beyond the beauty of the reef. Many countries depend on these fish for survival. Imagine widespread famine as fish stocks dwindle or move to cooler climates with no viable replacement.

Consider the UK and its traditional dish of fish and chips. Decades ago, cities like Hull and Grimsby had hundreds of fishing boats. Today, such industries lie in ruins, a resounding reminder of the results of overfishing and environmental degradation. Fish stocks move further north each year in search of cooler Arctic waters.

In the 1970s, the UK even went to war with Iceland—the Cod Wars—to protect its dwindling fish stocks. Iceland, determined to preserve its fishing grounds, declared a 300-mile exclusion zone, leading to clashes between fishing fleets and even a British warship being rammed.

All of this was not just a historical reference; I lived it. From the 1950s to the 1970s, the world witnessed the stubborn enmity between

Iceland and the UK over the rapidly contracting North Atlantic cod stock, accurately termed the "Cod Wars."

I was on a destroyer during that conflict. The tension in the atmosphere was as thick as Iceland's determination to protect its vital fishing industry, which took a bold stance against a much larger naval power. Their gunboats, dwarfed by our vessels, were outfitted with a desperate measure: concrete-filled bows. These makeshift battering rams were used to ram British fishing trawlers and naval frigates, one even sustaining significant damage.

This incident—the Icelandic gunboat Thor causing significant damage to a British Royal Navy frigate—was a turning point. British Prime Minister Margaret Thatcher, whose government I served under then, decided to send in reinforcements. Moreover, here is where I came in: My destroyer, the HMS Fife, a mighty guided-missile destroyer, was deployed to the region on that mission.

Iceland's navy, consisting of just two warships, was no match for our firepower. The sight of our imposing vessel likely deterred any further aggressive actions. The British government finally relented and agreed to Iceland's new limits to avoid risking the NATO agreement between both nations. The "Cod Wars" were perhaps just another grim reminder of how far nations would go to defend their resources and the catastrophic consequences of overfishing.

It is a vicious cycle: Climate change destroys marine ecosystems, leading to food shortages and increased demand for meat production. This, in turn, accelerates deforestation, further disrupting our planet's delicate balance.

***

I was a young man from a Staffordshire farming community who joined the navy and found myself in a conflict called the Cod Wars. This was not just about overfishing but a symptom of a more significant problem—a changing climate that was disrupting fish stocks. The term "global warming" was not widely used back then, but the effects were undeniable.

Fast forward to today, and the news is even more alarming. Recent articles discuss the potential collapse of the Gulf Stream, a massive ocean current that regulates Europe's weather. If this happens, Europe could plunge into a mini-ice age, devastating the climate and fish populations that sustain the continent.

The consequences would be dire. Food shortages would become a reality, with Europe relying heavily on imported food at skyrocketing prices. People would be forced to burn more fossil fuels for heat, exacerbating the very problem that caused the crisis.

This is not just a hypothetical scenario. The industrial fishing practices of some countries, particularly China and Japan, employ massive "trawling" ships that function as factory ships. They catch, process, and can the fish on the same boat. However, in their wake, they leave disaster in their path. Their practices harm the fish; they crush the entire marine ecosystem.

The urgency of climate action extends beyond environmental concerns. Water scarcity, a consequence of climate change, is already a flashpoint for global conflict. The situation in Kashmir is a sharp reminder of how competition for water can lead to violence. Under the effects of climate change, clashes over vital resources, including water, labor, soil, food, and energy, can easily break out.

The financial burden of climate change is already staggering and will only worsen. Rebuilding after natural disasters costs billions of dollars annually, a figure projected to balloon into the trillions. Who will foot the bill? Countries, including the United States and European powerhouses, will eventually reach their breaking points.

Taxing carbon polluters is a popular solution, but a significant catastrophe can wipe out any financial gains. Insurance companies, once considered a safety net, are starting to crumble. Florida serves as a cautionary tale. Hurricane Ian (2022) is the most recent tropical cyclone to rank in the top five costliest in US history, causing at least $114 billion in damage in the US. It is also the most expensive hurricane in Florida's history, surpassing Irma in 2017. Many residents who lost homes in hurricanes like these are still waiting for insurance payouts, with companies declaring bankruptcy and leaving the state altogether.

The incomprehensible insurance cost has tripled in just two years for most Florida residents. If they can afford it, they still need coverage. However, in many cases, most insurance companies don't even cover such disasters. In cases where they do, a bankruptcy case can quickly leave them high and dry.

Even the federal government wouldn't be immune as an ultimate backstop. With insurance companies collapsing, the government will be left holding the bag, a burden it simply won't be able to sustain. Reinsurance companies, once a safety net for regular insurers, are pulling out of high-risk areas like Florida, leaving a monopoly market with limited options.

The high insurance cost is leading some to take a gamble—self-insuring. These individuals choose to forgo insurance altogether, hoping to avoid the exorbitant premiums. It is a risky move, but the insurance cost has become prohibitive for some. This is only an option for those who own their property outright, with no mortgage or banking obligations.

This is not just a hypothetical scenario. Hurricanes like Katrina and Andrew have devastated entire regions; even recent storms have left lasting scars. Louisiana, particularly New Orleans, is still reeling from the damage. While relatively unscathed compared to Florida and Louisiana, New York is not immune to extreme weather events.

The reality is that these events are not a matter of "if" but "when." The rising frequency and intensity of natural disasters will disproportionately impact heavily populated areas, leading to widespread devastation and displacement. Relocating entire populations would be a logistical and financial nightmare, adding another layer of complexity to an already dire situation.

By acting now, we can prevent these financial catastrophes and safeguard the future for generations to come.

Demystifying Climate Science

Climate science studies long-term weather patterns and their influences, including natural variations and human activities (National Aeronautics and Space Administration, n.d.). Scientists use a diverse toolkit to understand climate:

Temperature Data: Ground-based weather stations, satellites, and ice cores extracted from glaciers provide historical and current temperature readings from land and sea, allowing scientists to identify trends and patterns (National Oceanic and Atmospheric Administration, 2023).

Proxy Records: These indirect indicators of past climate, such as tree rings, ocean sediments, and fossilized pollen, offer insights into

Earth's climate over vast periods (National Academies of Sciences, Engineering, and Medicine, 2016).

Climate Models: Complex computer simulations factor in variables like atmospheric conditions, ocean circulation, and solar activity. These models predict future climate scenarios based on different emissions trajectories (Intergovernmental Panel on Climate Change, 2021).

The Greenhouse Effect

Our planet relies on a natural phenomenon called the greenhouse effect to maintain a habitable temperature. Sunlight reaching Earth warms the surface, and some heat radiates back into space. Greenhouse gases, such as carbon dioxide and methane, act like a blanket, trapping some of this heat and preventing its escape. This natural process keeps Earth warm enough to sustain life (NASA, n.d.).

Human Influence

The issue arises when human activities significantly increase the concentration of greenhouse gases in the atmosphere. Burning fossil fuels like coal, oil, and natural gas releases large amounts of carbon dioxide. Deforestation removes trees that absorb carbon dioxide from the atmosphere. Replacing forests with methane-producing cattle farming only compounds the problem. These combined factors amplify the natural greenhouse effect, leading to global warming (NASA, n.d.).

The Evidence is Clear

The scientific consensus is overwhelming: human activity is the primary driver of climate change. The past decade has been the hottest on record, with rising sea levels, melting glaciers, and more extreme weather events pointing toward a warming planet (NASA, 2023).

However, the message is clear: climate change is real, human-caused, and a threat that demands our attention. Understanding the science allows us to move forward with solutions and build a more sustainable future.

Climate Change Models

Climate change models are potent tools scientists use to predict the potential impacts of our current course. These sophisticated computer simulations factor in variables like greenhouse gas emissions, ocean circulation, and atmospheric conditions. By running different scenarios, scientists can project how Earth's climate might change in the coming decades.

Imagine this: an eight-inch rise in sea level might seem insignificant. However, when combined with a powerful hurricane, the consequences for coastal cities like Miami or New York could be devastating. Climate models help us visualize these potential outcomes, urging us to consider the long-term implications of our actions.

The Looming Threat of Rising Seas

By the end of this chapter, you will have a comprehensive understanding of climate change, its causes, and its urgency. We have explored the science, delved into climate models, and examined the phenomenon's core – rising temperatures. Polar ice caps melt as temperatures rise, contributing to a significant rise in sea levels.

This poses a significant threat to humanity, considering that most of the world's population lives within a few hundred miles of coastlines. To put this into perspective, roughly 40% of the world's population lives within 100 kilometers (62 miles) of a coastline. This translates to an estimated 5.2 billion people potentially impacted by rising sea levels (Coastal Zone Management, n.d.).

Island Nations on the Brink

Many island nations face an existential threat from rising sea levels. The Maldives and Mauritius are just two examples. Some experts predict these idyllic island paradises could disappear entirely within a few years. Even smaller, lesser-known island nations might be completely submerged. Imagine an entire country's unique culture and history vanishing beneath the waves.

Coastal Cities Under Siege

The United Kingdom, for instance, has witnessed unprecedented flooding in recent years. Venice, the "City of Canals," is famously sinking as sea levels rise. This beautiful city, an art and history treasure trove, could be lost forever. These are just a few examples – the reality is that coastal communities worldwide are under siege from rising seas.

By understanding our challenges, we can work toward solutions and build a more resilient future.

A Call to Action

Kofi Annan's powerful words, "We are the first generation to be aware of the threat of climate change, and also the last generation that can take action to address it," are a stark reminder of our responsibility. We stand at a crossroads, holding the future of our planet in our hands.

This chapter has vividly illustrated the potential consequences of inaction: rising sea levels swallowing coastlines, extreme weather events wreaking havoc across the globe, and mass displacement due to water scarcity and food shortages. These are not distant threats; the impacts of climate change are already being felt worldwide.

However, there is still hope. We can chart a new course by acknowledging the situation's urgency and moving beyond the unproductive blame game. Climate change models, while not guaranteed, provide us with a glimpse into the future, a chance to course-correct before it's too late.

Breaking the Cycle of Blame

Pointing fingers – baby boomers accusing millennials, millennials blaming Gen X – achieves nothing. The truth is that climate change is a complex problem with roots in our collective industrial past. The baby boomer generation oversaw a period of rapid growth and development, and while the environmental impact was not fully understood then, it does not negate their accomplishments. They rebuilt a war-torn world and ushered in an era of relative peace and prosperity.

However, the world has changed. We, the inheritors of that progress, now understand our environmental challenges. Unlike previous generations, we cannot claim ignorance.

A Shared Future

The finger-pointing needs to stop. We are all in this together. With their wealth of experience, baby boomers can be powerful advocates for change. Brimming with innovation and fresh perspectives, millennials and Gen Xers can be the architects of new solutions.

Empowering a Sustainable Future

The solutions lie in a multi-pronged approach. Transitioning away from fossil fuels and embracing renewable energy sources is critical. Holding high-carbon-footprint industries accountable through taxation or stricter regulations can incentivize cleaner practices. Investing in sustainable technologies and infrastructure will pave the way for a greener future.

Individual actions, however small, can collectively create a significant ripple effect. Reducing our carbon footprint through conscious choices, from transportation to consumption habits, makes a difference. We can create a powerful movement by working together, demanding change from our leaders, and supporting sustainable businesses.

Kofi Annan's final words in that 2009 quote offer a glimmer of optimism: "We constantly underestimate the power of new technologies and overestimate the power of old institutions." Let us harness our generation's ingenuity to develop innovative solutions and empower new institutions that prioritize the well-being of our planet.

This is our defining challenge, and our choices today will determine the fate of future generations. Let us embrace responsibility and build a harmonious future where humanity and nature thrive.

CHAPTER 2
Legacy In The Balance: The Boomers' Climate Quest

Baby boomers, born roughly between the mid-1940s and the mid-1960s, have a complex relationship with climate change. A recent study shows that people over 60 were responsible for nearly a third of greenhouse gas emissions in 2015 (Barnes, 2022). However, this still needs to negate the fact that baby boomers can still be a powerful force for positive environmental change.

As a member of the baby boomer generation (born in 1959), I am keenly aware of this responsibility. The study highlights a sobering truth: 80% of carbon emissions occurred after World War II (Crippa et al., 2022). This period coincides with the rise of consumerism and economic growth that many boomers benefited from. We must acknowledge our role in creating the climate crisis we face today.

However, acknowledging the problem is only the first step. Baby boomers have the time, energy, and financial resources to invest in renewable energy solutions, support climate-friendly policies, and mentor younger generations taking action on climate change.

This book will explore how baby boomers can use their experiences and resources to leave a positive environmental legacy. We will examine specific actions boomers can take, from adopting sustainable practices in their own lives to advocating for climate action at the community and national levels. By taking responsibility for the past and actively working toward a greener future, baby boomers can ensure their legacy is one of environmental stewardship.

While the Industrial Revolution undoubtedly contributed to the environmental problems we face today, it is essential to recognize the significant impact of the post-war baby boomer generation on climate change. As a British-born baby boomer, I can vividly recall the post-war era in the UK. The country, a leader in the Industrial Revolution, relied heavily on coal for electricity, railways, and various industries like textiles, steel, and shipbuilding.

However, it is crucial to remember that much of that heavy industry has since disappeared. Once a leader in industrial pollution, the UK has taken steps toward a cleaner future. Birmingham, my birthplace, exemplifies this shift. Although it remains the UK's second-largest city, it is no longer synonymous with heavy industry and air pollution.

The baby boomer generation's impact on climate change is undeniable. Meland (2022) highlights our significant contribution to greenhouse gas emissions. It is easy to understand why – environmental consciousness was not a prevalent concern during our formative years.

The post-war era was a time of survival and rebuilding. Countries focused on economic recovery, feeding their populations, and providing housing for returning soldiers. Women, who traditionally stayed home, began entering the workforce in more significant numbers during this period. This societal shift, coupled with the latter part of the 1960s and the sexual revolution, significantly altered consumption patterns. The increased participation of women in the workforce was not just about individual aspirations; it was a critical factor in national economic growth. Recognizing the value of this "human capital" was essential for post-war recovery. This newfound economic power fueled a desire for new cars, homes, appliances, and other goods, contributing to a rise in greenhouse gas emissions.

However, acknowledging our role in the problem is just the first step. Baby boomers have the potential to be a powerful force for positive change.

Baby boomers present a complex picture regarding climate change. They have a mix of proactive and empathetic attitudes alongside a population segment that remains unconcerned.

Many baby boomers recognize the environmental impact of their generation's consumerism and industrial practices. They feel a sense of responsibility for the current climate crisis, and this acknowledgment often fuels a desire to take action.

An interesting parallel exists between the public's evolving view on climate change and past societal shifts regarding health concerns. Just as many boomers grew up in a time when smoking was not only acceptable but also promoted, the long-term consequences of environmental damage were not widely understood.

The analogy extends further. Just as public health campaigns successfully transformed societal attitudes toward smoking, similar campaigns can raise awareness about climate change. The key lies in creating a cultural shift emphasizing the importance of environmental responsibility.

Economic Considerations and the Power of Alternatives

However, acknowledging the need for change does not automatically translate to action, especially when economic factors are involved. For example, the dairy industry, heavily reliant on marketing campaigns promoting milk as an essential source of calcium, has created a perception that plant-based alternatives are less economical.

This highlights the importance of creating readily available and affordable alternatives. Consumers will likely embrace a sustainable choice if plant-based milk options can achieve price parity with dairy milk. Achieving this shift requires investment in sustainable alternatives and potentially re-evaluating subsidies that favor specific industries.

It is essential to acknowledge a generational gap regarding climate change concerns. Studies like the Pew Research Center consistently show that younger generations are more concerned about climate change than their baby boomer counterparts.

For example, a 2021 Pew Research Center study found that 67% of Generation Z (born 1997-2012) believes climate change should be a top priority, compared to only 57% of baby boomers and those older. Similarly, younger generations are more likely to engage in social media activism and support climate-friendly policies.

The Power of Market Forces and Shifting Consumer Preferences

The dairy industry's example highlights the powerful influence of market forces in shaping consumer behavior. Decades of marketing campaigns promoting dairy products as essential for health generated a robust cultural association. However, this narrative is shifting as consumers become more informed about the environmental impact of animal agriculture and the availability of healthy plant-based alternatives.

The key to a successful transition is making these alternatives accessible and appealing. If plant-based milk options can achieve price parity with dairy milk, the economic incentive to switch becomes significant. This does not necessarily require persuading the end user – the convenience and affordability of the sustainable option will naturally drive market forces.

This is where baby boomers have a unique opportunity to influence change. Many boomers hold positions of financial power through personal investments or control over investment portfolios. By strategically directing investments toward sustainable businesses and environmentally friendly practices, they can send a powerful message to the market.

Furthermore, boomers can influence their political impact by advocating for policies that promote clean energy and responsible consumption. Their voices carry weight with policymakers, and their support for these initiatives can drive legislative change.

The transformation of the tobacco industry offers a valuable parallel. For decades, aggressive marketing campaigns normalized cigarette smoking. However, a growing body of scientific evidence and public health campaigns advocating for the dangers of smoking led to a significant shift in social norms. Regulations on advertising and increased public awareness played a crucial role in curbing tobacco use.

Similarly, addressing climate change necessitates a multi-pronged approach. While public education and awareness campaigns are crucial, regulations on environmentally harmful practices and incentives for sustainable alternatives can significantly accelerate progress.

As the baby boomer generation ages, they will face a unique set of challenges posed by climate change. Extreme weather events, including heat waves and air pollution, will disproportionately impact older adults. Many boomers already grapple with chronic health conditions such as heart disease, diabetes, and respiratory problems. These existing vulnerabilities make them more susceptible to the health risks associated with rising temperatures and declining air quality.

Baby boomers hold significant political and economic power. They are a dominant voting bloc with the ability to influence policy decisions. Many boomers hold leadership positions in businesses and industries. By strategically using their influence, they can advocate for policies that promote clean energy, sustainable practices, and climate resilience.

Furthermore, boomers can use their financial clout to support businesses committed to environmental responsibility. Divesting fossil fuel companies and investing in renewable energy initiatives sends a powerful message to the market, and this financial leverage can be a powerful tool for driving positive change.

While it is true that addressing climate change may require expensive solutions, such as installing solar panels or purchasing electric vehicles, focusing on practical, everyday actions can make a significant difference. Simple steps like combining errands, carpooling, walking or cycling for short trips, and managing water and energy consumption can contribute to a reduced carbon footprint.

Electric bikes are a fantastic option for local trips. Their price has decreased significantly recently, with some models costing less than $600. These bikes boost pedaling, making it easier to cover longer distances or navigate hilly terrain. Many electric bikes can be charged in just a few hours, making them a convenient and eco-friendly choice for everyday errands and commutes.

Sustainable Building Practices

The choices we make about building materials have a significant impact on the environment. Baby boomers looking to reduce their environmental footprint can consider incorporating sustainable building practices into home renovation or construction projects.

Here are some eco-friendly materials to consider:

Using recycled glass, plastic, or metal materials reduces demand for virgin resources and lessens the environmental impact of production.

Look for wood with certifications like FSC (Forest Stewardship Council) to ensure it comes from responsibly managed forests.

New technologies are emerging to reduce the carbon footprint of concrete production.

Giving new life to old materials like bricks, doors, or beams saves resources and adds character to a building project.

Bamboo is a fast-growing and renewable resource that offers a beautiful and sustainable alternative to traditional hardwoods.

Using recycled steel in construction projects reduces reliance on virgin materials and lowers energy consumption during production.

By incorporating these sustainable practices, baby boomers can build or renovate beautiful, functional, and environmentally responsible homes. These actions benefit the environment and lead to cost savings. By highlighting the financial benefits of sustainable practices, we can encourage broader participation, even among those with limited resources.

Dietary Choices and Reducing Environmental Impact

Another area where boomers can make a difference is in their dietary choices. Animal agriculture, particularly beef production, has a significant environmental impact. For those who occasionally enjoy meat, selecting "green meat" options raised with sustainable practices can be a more environmentally friendly choice.

The production of one beef burger can generate significant greenhouse gas emissions. According to a study by Consumer Ecology, a typical 4-ounce beef burger has a carbon footprint of 9.73 kg CO2e. In comparison, beef's greenhouse gas emissions are 24.5 times higher than soy's. Consuming just 4 ounces daily of red meat contributes to a land footprint of 112,881 square feet annually, equivalent to 784 parking spaces.

Reducing red meat consumption and opting for plant-based alternatives is a simple yet impactful action individuals can take. This shift aligns with healthy eating recommendations, offering a win-win for personal health and the environment.

Simple Solutions and Everyday Impact

Beyond large-scale initiatives, there are numerous everyday actions baby boomers can take to reduce their environmental footprint. Refusing single-use plastic bags at grocery stores is a simple yet impactful practice. Many stores offer reusable shopping bags as an alternative, and some even provide paper bag options that are more biodegradable than plastic (American Forests, 2023). These small changes can significantly impact plastic waste reduction when compounded across a large population.

Similarly, making small changes at home can significantly reduce energy consumption. Here are a few examples:

Install a timer on your water heater: This appliance can be a significant energy consumer. Setting a timer to ensure it only heats water when needed can lead to substantial cost savings and reduced environmental impact. Studies show that water heating can account for hundreds or even thousands of dollars on your energy bill annually.

Consider using an air fryer instead of a large oven for smaller meals. Air fryers are becoming increasingly popular due to their energy efficiency. They can consume significantly less energy for smaller meals than a traditional oven.

Think twice about running the dishwasher: Dishwashers are convenient but energy-intensive. If you only have a half-load, consider washing dishes by hand instead. When you do use the dishwasher, opt for an energy-efficient model and choose the air-dry setting whenever possible.

Estate Planning and Environmental Legacy

Estate planning, outlining how assets will be distributed after death, presents a unique opportunity for baby boomers to incorporate environmental considerations. Charitable giving to organizations focused on clean energy, sustainable practices, and ecological conservation allows boomers to leave a legacy that benefits the planet (Land Trust Alliance, 2023).

Extending the Life Cycle of Sustainable Assets

Another way to minimize your environmental footprint is by choosing a home with easy access to daily necessities. Look for properties located near grocery stores, schools, and healthcare facilities. This reduces reliance on cars for errands and allows for walking, cycling, or public transportation whenever possible.

Solar panels, a significant investment for many homeowners, typically have a lifespan of around 25 years. However, advancements in technology are constantly improving efficiency. Baby boomers who have installed these new solar panels can consider donating their older systems to individuals or organizations who may need help to afford a brand-new installation. This approach extends the life cycle of these sustainable systems and allows others to benefit from clean energy.

The concept of environmental stewardship extends beyond individual assets like solar panels. Baby boomers passing down homes or vehicles can emphasize the importance of using these assets in an eco-conscious manner. An energy-efficient house with water-saving devices becomes a model for sustainable living for future generations. Similarly, passing on an electric vehicle encourages continued use of clean transportation.

However, the responsibility for environmental stewardship continues beyond simply passing down assets. Baby boomers can educate their heirs about the ecological impact of their choices. Encouraging them to invest in sustainable businesses and adopt eco-friendly practices in their daily lives ensures that the legacy extends beyond the material. Discouraging them from indulging in gas-guzzling vehicles, even high-performance electric models with artificial engine sounds, reinforces the commitment to a sustainable future.

For example, Alfa Romeo, a renowned Italian sports car brand, has launched a powerful 1000-horsepower electric car. While impressive in performance, baby boomers can use this to educate younger generations about responsible choices within the electric vehicle market. Electric vehicles are available that prioritize efficiency and environmental impact without sacrificing the driving experience.

The Ripple Effect

Hurricane Maria, which ravaged Puerto Rico in 2017, serves as a stark reminder of the devastating consequences of climate change. The powerful storm caused widespread destruction, leaving millions without power and access to clean water for months. In the aftermath of this disaster, a wave of climate activism emerged, showcasing the firm spirit of those on the front lines of the fight for a sustainable future.

Sixteen-year-old climate activist Greta Thunberg is a powerful symbol of this movement. Her now-iconic school strike for climate action in 2018 inspired millions worldwide to demand action from world leaders. Thunberg has since been arrested multiple times for participating in peaceful protests, demonstrating the lengths to which young activists are willing to go to raise awareness about the climate crisis.

Beyond Thunberg, countless individuals and organizations work tirelessly to address climate change. From local grassroots movements to international environmental NGOs, these champions advocate for clean energy policies, promote sustainable practices, and hold corporations accountable for their environmental impact.

Creating an Eco-Friendly Estate Plan

Estate planning is not just about distributing assets; it is an opportunity to leave a lasting legacy. Baby boomers can incorporate environmental stewardship into their estate plans by considering the following:

Green Investments: Direct investments in renewable energy companies or funds that support sustainable practices demonstrate a commitment to a greener future.

Charitable Giving: Numerous environmental organizations are working on critical issues like deforestation, clean water access, and conservation. Donating to these organizations allows your legacy to support these crucial efforts directly.

Sustainable Property Management: Consider the environmental footprint when passing down the property to heirs. Homes with solar panels, energy-efficient appliances, and water-saving features benefit the environment and reduce utility costs for future generations.

Open Communication: Engaging in open conversations with heirs about the importance of environmental responsibility ensures that inherited assets are used sustainably. Discuss the benefits of electric vehicles, renewable energy sources, and responsible consumption habits.

In conclusion, baby boomers can leverage their experience, resources, and influence to combat climate change. By acknowledging their generation's role in the crisis, adopting sustainable practices in their daily lives, advocating for climate-friendly policies, and incorporating environmental considerations into their estate plans, baby boomers can leave a legacy of environmental responsibility for future generations.

CHAPTER 3
Generation X: The Bridge Of Change

A report by BC Hydro (2022) highlights an interesting aspect of climate action: Generation X, the bridge between baby boomers and millennials. This chapter explores Gen X's pivotal role in tackling the climate crisis.

While the report notes that Gen Xers (aged 42-57) tend to live in larger homes with a higher environmental footprint, this generation also holds a unique position. They witnessed firsthand the dawn of the technological and economic shifts of the late 20th and early 21st centuries.

Gen X was at the forefront of the personal computer revolution, the rise of the internet, and the burgeoning green economy. These advancements offered tools to mitigate climate change. Gen X witnessed and participated in these changes, from developing renewable energy technologies to developing sustainable practices.

Their careers and lifestyles adapted to these advancements, making them conduits to the problem and potential solutions.

As they entered their late 30s and 40s, Gen X began to grapple with the stark realities of climate change. Unlike baby boomers, who may not have witnessed the environmental consequences as readily, Gen X saw the tangible effects on the planet. This realization sparked a sense of urgency and a commitment to finding solutions.

Furthermore, Gen Xers are at a stage in life where many have accumulated financial resources. This economic power allows them to become significant changemakers.

Leading by Example

Several prominent Gen X figures exemplify this commitment to sustainability. Sunita Narain, a renowned Indian environmentalist and activist (Wikipedia, 2024), showcases the global reach of Gen X's climate leadership. These individuals, along with countless others, demonstrate the power this generation holds in driving positive change.

Bridging the Divide

Gen X occupies a unique space between baby boomers, who may not have fully grasped the environmental consequences of their actions, and millennials, who grew up acutely aware of the climate crisis. This positioning allows Gen X to bridge the gap by fostering understanding and collaboration.

Through their lived experiences and accumulated knowledge, Gen X can mentor younger generations and translate environmental awareness into concrete action. Furthermore, their financial resources and established careers enable them to influence policy changes and support sustainable business practices.

Addressing Traditional Practices

While Gen X plays a crucial role in bridging the sustainability gap, tackling climate change requires a multifaceted approach that considers cultural and religious practices across the globe.

A prime example lies in open-air cremation, a tradition prevalent in some parts of the world, particularly in India. The traditional method of cremating bodies on open pyres, often along the banks of the Ganges River in Varanasi, India, carries a significant environmental burden. Wood, a critical component of this process, is a finite resource, and its unsustainable harvesting contributes to deforestation (Hindustan Times, 2021).

The incomplete combustion associated with open-air cremation releases harmful pollutants into the atmosphere, contributing to smog and air quality issues (Central Pollution Control Board, India, 2023).

Addressing this issue requires sensitivity and respect for deeply rooted cultural and religious beliefs. Simply condemning the practice is counterproductive. Instead, the focus should be on fostering dialogue and exploring alternative solutions that are environmentally conscious and respectful of tradition.

One approach could involve posing open-ended questions that encourage reflection within communities, such as:

1. "Given environmental concerns, can we explore ways to conduct cremation ceremonies that are more sustainable while still honoring religious traditions?"
2. "Are there alternative methods of cremation that could reduce the environmental impact without compromising spiritual significance?"

This approach avoids imposing solutions but encourages communities to converse about potential adaptations. This way, we could explore existing and emerging sustainable cremation options, such as:

Electric Cremation: This method uses electricity to cremate bodies, eliminating the need for wood and significantly reducing air pollution (BBC News, 2023).

Bio Cremation: This eco-friendly technique utilizes a natural, alkaline solution to decompose the body, resulting in a smaller volume of cremated remains (Resomation Ltd., 2024).

Presenting these alternatives and fostering dialogue can bridge the gap between environmental concerns and cultural traditions. Collaboration with religious leaders, environmental organizations, and local communities can play a vital role in developing and implementing sustainable solutions.

Open-air cremation highlights the complex interplay between environmental sustainability and cultural practices. Finding solutions requires a delicate balance between respecting traditions and protecting the planet. Gen X can play a crucial role in achieving this balance by facilitating respectful dialogue and exploring sustainable alternatives.

Gen X and the Legacy of Environmental Legislation

The narrative of Gen X and climate change extends beyond their consumption habits. This generation occupies a unique historical position, having witnessed significant environmental legislative advancements in the late 20th century.

A Legacy of Clean Air

One crucial example is the 1990 amendments to the Clean Air Act of 1970 (Environmental Protection Agency, 2024). These amendments tackled critical issues like acid rain, toxic emissions, and ozone depletion. Gen X came of age during this period, witnessing firsthand the power of legislation to address environmental challenges.

Global Initiatives

Beyond the United States, Gen X witnessed the development of groundbreaking international initiatives. The European Emissions Trading Scheme, a cornerstone of the EU's climate change strategy (European Commission, 2023), and Germany's pioneering Renewable Energy Act (BMU Bundesministerium für Umwelt, Naturschutz und Reaktorsicherheit, 2021) are prime examples. These developments further solidified the growing global focus on environmental sustainability.

Bridging the Divide Through Sustainable Practices

While Gen X may face immediate pressures like financial stability and family responsibilities, they can still play a significant role in fostering a sustainable future. One strategy is "sustainable parenting," where Gen Xers pass their environmental awareness to the next generation. This can involve practical tips on resource conservation, responsible consumption, and environmentally conscious living. Here are some ways Gen X parents can embrace sustainable practices and raise eco-conscious children:

Connect with other green parents: Share tips, swap resources, and organize eco-friendly playdates. You can create a supportive network that reinforces sustainable values for your children.

Embrace sustainable laundry: Choose cold washes, eco-friendly detergents, and air drying whenever possible. Let your kids help hang clothes on a line, teaching them about resource conservation.

Switch to greener cleaners: Ditch harsh chemicals and adopt natural cleaning solutions like vinegar and baking soda. This will protect your family's health and minimize environmental impact.

Reduce car trips: Walk, bike, or use public transportation whenever possible. This reduces emissions and encourages physical activity for the whole family.

Eat sustainably: Plan your meals, buy local and seasonal produce, and reduce food waste. Involve your kids in meal planning and composting scraps, fostering an appreciation for responsible consumption.

Be an example: Your children are constantly observing your behavior. Make conscious choices about energy usage, recycling, and waste reduction. Let them see your commitment to a greener lifestyle.

Open a dialogue: Talk to your kids about environmental issues in an age-appropriate way. Please encourage them to ask questions and express their concerns.

Get them outdoors: Foster a love for nature by taking your children on hikes, exploring parks, or simply spending time in your backyard. These experiences cultivate respect for the environment.

Nurture a connection with nature: Teach your children to care for plants and animals. Start a small garden together, volunteer at an animal shelter, or build a bird feeder. These activities instill a sense of responsibility and empathy for the natural world.

Upcycle: Teach your children about waste management. Teach them to reduce, reuse, and recycle. Get creative with upcycling projects—transform old toys into new ones or turn used jars into planters.

Volunteer together: Give back to your community by volunteering for environmental causes. This provides valuable real-world experience and empowers your children to make a positive difference.

By incorporating these practices into your family life, you raise environmentally conscious children and foster a generation of responsible stewards for our planet. Remember, there is always time to start. Take action today and embark on this exciting journey toward a sustainable future with your children by your side.

Influencing Corporate Sustainability

Furthermore, Gen X occupies influential positions within corporations and businesses. Their lived experiences with environmental legislation and growing awareness of climate change can empower them to advocate for sustainable practices within their organizations. They can champion green initiatives, promote eco-friendly policies, and hold businesses accountable for their environmental impact. Here is how Gen X can lead the charge:

Measurable Goals for Sustainability: Gen X can advocate for establishing clear, measurable environmental goals. This could involve setting reduced paper usage, energy consumption, or waste generation targets. By quantifying progress, companies can track their impact and celebrate achievements.

Going Paperless: Gen X can spearhead a shift toward a paperless workplace. Encourage digital document storage, double-sided printing, and online forms to reduce paper waste and save resources.

Combating Vampire Power: Gen X can champion the fight against "vampire power" – the energy electronic devices continue to draw even when supposedly off. Educate colleagues on powering down computers, monitors, and printers at the end of the day.

Indoor and Outdoor Plants: Gen X can advocate for incorporating plants throughout the office. Plants beautify the workspace and improve air quality by filtering out pollutants.

Lighting the Way with Efficiency: Gen X can push to replace traditional bulbs with energy-efficient LEDs. This reduces energy consumption and saves the company money on electricity bills.

Turning Off Lights: Gen X can lead by example, turning off lights when leaving a room or during breaks. Encourage colleagues to adopt this habit, reducing unnecessary energy waste.

Recycling Right: Gen X can spearhead a robust recycling program in the office. Clearly labeled bins and educational campaigns ensure proper waste segregation and diversion from landfills.

Donation Programs: Gen X can initiate donation programs for unused office supplies or furniture. This reduces waste and allows these items to find a second life elsewhere.

The Single-Use Habit: Gen X can advocate for reducing single-use plastics and disposables. Encourage using reusable water bottles, coffee mugs, and lunch containers.

Green Events for a Greener Future: Gen X can organize green events to raise awareness about sustainability. This could involve hosting a clothing swap, a farmers market visit, or a tree-planting initiative.

The Power of Remote Work: Gen X can advocate for remote work options when feasible. This reduces employee commutes, leading to fewer vehicles on the road and lower carbon emissions.

Sustainable Packaging: Gen X can push for using eco-friendly packaging materials for company products. This could involve switching to recycled cardboard, biodegradable materials, or minimal packaging.

Supporting Green Suppliers: Gen X can encourage the company to source materials and services from environmentally responsible suppliers. This creates a ripple effect, supporting businesses committed to sustainability.

Exploring Alternative Energy: Gen X can champion exploring alternative energy sources like solar or wind power. This reduces reliance on traditional, often polluting, energy sources. Gen X can advocate for installing solar panels in company buildings, leading to significant long-term cost savings on electricity bills while producing clean energy.

By championing these initiatives, Gen X can transform workplaces into sustainability models. Their leadership and commitment will pave the way for a greener future for all.

Gen X's Role in Climate Action

Gen X's unique position between the baby boomer and millennial generations makes them a crucial bridge in the fight against climate change. Their experiences, perspectives, and financial resources empower them to connect these often disparate groups and drive meaningful action.

Understanding the Divide

Baby boomers, who came of age during economic prosperity and technological advancements, may have yet to grasp the long-term environmental consequences of their actions thoroughly. On the other hand, millennials grew up acutely aware of climate change and its urgency. This generational gap can lead to misunderstandings and a lack of collaboration, hindering progress.

Gen X's Bridge-Building Potential

Gen X witnessed the environmental movement gain momentum in the late 20th century. They saw the passage of landmark legislation like the 1990 amendments to the Clean Air Act (Environmental Protection Agency, 2024) and the rise of international initiatives like the European Emissions Trading Scheme (European Commission, 2023). This firsthand experience positions Gen X to translate environmental awareness into concrete action and bridge the gap between boomers who may need more convincing and millennials eager for immediate change.

Fostering Dialogue and Collaboration

Gen X can facilitate dialogue between generations by highlighting the shared goal of a sustainable future for all. Open-ended questions posed by Gen X can encourage boomers to reflect on the environmental impact of their choices and millennials to consider practical solutions. Collaborative workshops, community events, and intergenerational mentoring programs can foster understanding and collaboration.

Financial Power for Change

Many Gen Xers are now at a stage in life where they have accumulated financial resources. This economic power allows them to contribute significantly to climate action. They can invest in sustainable businesses, support climate-friendly initiatives, and advocate for policies that prioritize environmental protection.

Several prominent figures within Gen X are potent examples of climate change leadership. Elon Musk, a tech entrepreneur driving innovation in electric vehicles with Tesla, exemplifies this commitment (CNBC, 2023). Jeff Bezos, founder of Amazon, has pledged billions of dollars through the Earth Fund to protect nature (Amazon Sustainability, 2023). Bill Gates, a tech pioneer, is actively tackling the global water crisis (Bill & Melinda Gates Foundation, 2024). These individuals, along with countless others, demonstrate the transformative potential of Gen X in the fight for a sustainable future.

Real-Time Witnessing of Climate Change

Gen X is the first generation to witness climate-driven disasters unfold in real-time through news media. Hurricanes, tornadoes, wildfires, and floods are no longer distant events reported in newspapers but immediate crises showcased across screens. This visceral experience fosters a deep understanding of the human cost of climate change, motivating Gen X to seek solutions and advocate for preventative measures.

Gen X is the bridge between boomers and millennials and the generation that will care for the aging boomer population. As grandparents, they have a unique opportunity to instill environmental values in the next generation. By adopting sustainable practices in their own lives and educating their grandchildren about the importance of a healthy planet, Gen X can leave a lasting legacy of environmental stewardship.

Ili Nadiah Dzulfakar, co-founder and chairperson of Klima Action Malaysia, cannot shy away from a challenge. Witnessing the rapid transformation of Malaysia's landscape firsthand - lush rainforests and peatlands devoured by monoculture plantations - sparked a fire within her. The environmental consequences were undeniable: biodiversity collapse, rising carbon emissions, and devastating floods brought on by a warming climate.

Nadiah was more than just a passive observer. Growing up in the '90s, she saw the Sumatran Rhino and the leatherback turtle vanish from Malaysia within her lifetime, a stark reminder of the ecological crisis unfolding. The near extinction of the Malayan tiger painted an even bleaker picture.

Driven by a sense of urgency and a desire for justice, Nadiah co-founded Klima Action Malaysia. This grassroots group prioritizes the voices of vulnerable communities most impacted by climate change—Indigenous communities, women, and young people. Its mission is to empower these groups and build a more inclusive climate movement.

Klima Action Malaysia fights for a just transition that acknowledges the historical injustices indigenous communities face. These communities, the guardians of Malaysia's remaining forests, have been denied sovereignty over their customary territories, and their knowledge of land management has been disregarded.

Nadiah's background perfectly positions her to lead this fight. Her experience working with Indigenous communities and her academic training in environmental science allows her to bridge the gap between traditional knowledge and scientific solutions.

Klima Action Malaysia does not just raise awareness; it actively empowers communities. They provide a platform for Indigenous women and youth to be recognized as crucial stakeholders in crafting solutions to the climate crisis. Their fight for a climate emergency declaration in Malaysia is a testament to their unwavering commitment to a sustainable future.

Nadiah's story is a beacon of hope in an environmental crisis. It exemplifies the courage and dedication of Malaysia's climate movement, which prioritizes inclusivity, justice, and the voices of those most affected.

Gen X is pivotal in bridging the generational divide on climate change; their experiences, financial resources, and leadership potential position them as powerful advocates for a sustainable future. By fostering collaboration, leading by example, and nurturing environmental awareness in future generations, Gen X can ensure a healthier planet for all.

The Ripple Effects of Conflict

While Gen X acts as a bridge between generations on climate action, the world around them continues to face complex challenges. The rise of Gen Z and the ongoing threat of geopolitical conflict necessitate a broader conversation about the interconnectedness of environmental issues, war, and economic disparity.

Millennials, unlike Gen X, were raised in a world shaped by continuous military interventions like the Gulf War, the Iraq War, and the conflict in Afghanistan (History Channel, 2024)-- this exposure to war and its consequences instilled in them a strong desire for peace and global stability. However, the current geopolitical climate, with tensions rising between the United States, China, and Russia, and the ongoing war in Ukraine, presents a stark reality check.

Beyond the human cost, war carries an enormous economic burden. Billions of dollars are funneled into weaponry, ammunition, and military operations, diverting resources from crucial areas like climate action and poverty alleviation. The destruction of infrastructure, displacement of populations, and disruption of trade further exacerbate economic hardship and fuel resentment.

The Root Causes of Conflict

A deeper examination reveals that economic disparity and resource scarcity are often at the root of conflict. Countries with limited economic opportunities and insufficient resources to meet their populations' basic needs become breeding grounds for instability. Addressing these underlying issues requires shifting focus from military solutions to investments in sustainable development, infrastructure projects, and job creation.

Investing in renewable energy infrastructure, sustainable agriculture, and water management in developing countries fosters self-sufficiency and reduces dependence on volatile resource markets. These measures empower communities, alleviate poverty, and create economic opportunities, ultimately promoting peace and stability.

The Deadliest Threat: Air Pollution

It is essential to emphasize that poor air quality linked to pollution from wars and industrial activities has claimed far more lives than wartime casualties (World Health Organization, 2023). This stark reality highlights the urgency of addressing climate change and transitioning to clean energy sources.

Land and Water Management

One solution could be building dams and irrigation systems in arid regions. These projects can provide much-needed access to clean water for drinking and agriculture, enabling communities to grow food and achieve self-sufficiency. This approach tackles water scarcity and economic insecurity, potentially reducing tensions and promoting peace.

The fight against climate change cannot exist in a vacuum. It is intricately linked to geopolitical tensions and economic disparity. With their heightened awareness of conflict and global issues, millennials can be a powerful force for change. By advocating for sustainable development, promoting peacebuilding initiatives, and demanding accountability from world leaders, they can work toward a future where environmental security and economic stability are cornerstones of a peaceful and prosperous world.

The Rise of the Global Workforce and Climate Action

The narrative around climate action often overlooks the transformative power of technological advancements like the internet and artificial intelligence (AI). So, how have these forces facilitated the rise of a global workforce and its potential impact on environmental sustainability?

Bangladesh serves as a compelling case study. Once heavily reliant on the textile industry, the country is witnessing a freelance work boom. Platforms like Upwork and Guru connect skilled Bangladeshi freelancers with clients worldwide, contributing billions of dollars to the national economy (Freelancers Bangladesh, n.d.).

The internet has shrunk geographical boundaries, creating a "global village" where location is no longer a barrier to employment. Individuals with the necessary skills can now secure jobs with companies located worldwide. This shift disrupts the traditional notion of work-life balance, allowing individuals to prioritize environmental consciousness without sacrificing economic security.

The Disappearing Commute

The rise of remote work offers a significant environmental benefit: reduced reliance on personal vehicles. Millions of daily commutes are no longer necessary, leading to a decrease in greenhouse gas emissions and air pollution. Companies that adopt remote work models can further contribute by reducing their office space needs and minimizing energy consumption for lighting, heating, and cooling.

Beyond Excuses: Balancing Work and Sustainability

The argument that economic necessity trumps environmental concerns becomes less valid in this new reality. Individuals can pursue a comfortable living and contribute to a sustainable future simultaneously. Technology empowers them to work remotely, eliminating the environmental cost of commuting.

The Case of India

India's booming software export industry provides another example. Skilled professionals in Bangalore and other Indian cities earn competitive wages while working remotely for international companies. This reduces their carbon footprint while contributing significantly to the Indian economy (TheCivilEngineer.org, 2024).

Remote Work's Unexpected Benefit

The COVID-19 pandemic, a global crisis with devastating consequences, inadvertently demonstrated the viability of remote work. Many businesses discovered that employees working from home could maintain or even increase productivity. This realization and environmental benefits have prompted many companies to adopt permanent remote work models.

While the pandemic was a tragedy, the shift to remote work it spurred can be seen as a "blessing in disguise." Companies are discovering a more efficient and environmentally friendly way of operating, and employees are gaining greater flexibility and control over their work environment.

Reflect, Refine, Reimagine

Gen X, you hold the key to unlocking a sustainable future. While you have witnessed environmental changes firsthand, the fight against climate change is far from over. Here are some prompts to spark self-reflection and inspire further action:

1. Beyond personal habits, how can you leverage your professional skills and experience to advocate for sustainable practices within your company?
2. Reflecting on your financial stability, explore ways to invest in or support businesses driving climate solutions.
3. Given your life stage, how can you mentor younger generations to become environmentally conscious leaders?
4. Living between boomers and millennials, can you brainstorm ways to bridge the generational divide on climate action?
5. Revisit your community involvement. Is there a local environmental cause you could lend your time and expertise to?

Conclusion

In conclusion, the BC Hydro report paints a compelling picture of Gen X's potential as a powerful force for positive environmental change. While their current lifestyles may be less eco-friendly, Gen X's unique experiences position them perfectly to bridge the gap between the past and the future. They witnessed the dawn of the green movement and the rise of environmental legislation firsthand, equipping them with the knowledge and tools to advocate for a sustainable future.

The key takeaway is that Gen X is the bridge between generations on climate change. Their financial resources, leadership potential, and lived experiences empower them to influence policy changes, promote sustainable practices within corporations, and inspire future generations.

Beyond simply acting as a bridge, Gen X has the potential to be the architect of a sustainable future. Their role as parents allows them to instill environmental values in their children, ensuring a future generation of responsible stewards for the planet. This chapter emphasized how Gen X can contribute to a greener future, from influencing corporate sustainability efforts to practicing sustainable parenting.

The fight against climate change, however, is a multi-generational effort. The next chapter will delve into the views and perspectives of millennials, exploring their unique contributions to the fight for a sustainable future. As you continue your journey toward a greener lifestyle, remember the power of Gen X. Embrace sustainable parenting practices, advocate for change in your workplace, and draw inspiration from the stories of those paving the way for a healthier planet.

CHAPTER 4
Millennials: Modernizing The Green Revolution

Greta Thunberg's quote powerfully opens this chapter's discussion on millennials' role in modernizing the green revolution. "The climate crisis is both the easiest and the hardest issue we have ever faced. The easiest thing is because we know what we must do. We must stop the emissions of greenhouse gases. The hardest because our current economics are still totally dependent on burning fossil fuels, thereby destroying ecosystems to create everlasting economic growth." (Speech to the British Parliament - April 23, 2019, 2019)

Thunberg argues that climate change presents a unique challenge. We understand the solution (curbing greenhouse gas emissions), but our economic dependence on fossil fuels creates a significant hurdle.

This chapter explores how millennials spearhead a digital-driven approach to climate activism, primarily through social media. Social media platforms leverage the power of networks, allowing a message to spread globally with unprecedented ease. Take Greta Thunberg, a young Swedish activist. Thanks to social media, her message has transcended geographic boundaries, reaching millions worldwide, including the United States (National Geographic, 2019).

This phenomenon is driven by the network effect, which describes how a product or service becomes more valuable as more people use it. In the context of social media and climate activism, each user sharing climate-related content expands the network and raises awareness. This digital movement empowers individuals to connect, share ideas, and mobilize large-scale action.

It is essential to acknowledge that a dedicated chapter will explore the technological aspects of this movement more deeply. However, this chapter lays the groundwork by highlighting social media's crucial role in amplifying voices and fostering a global community of climate activists.

Social media platforms empower geographically distant individuals to connect and share ideas, propelling a teenager's voice onto the international stage. This phenomenon highlights social media's unprecedented reach and impact on activism (Boykoff & Yulsman, 2013).

The discussion of climate change has traditionally been dominated by highlighting the gravity of the problem – greenhouse gas emissions and their detrimental impact on ecosystems.

This chapter proposes a shift in perspective. While raising awareness remains crucial, millennials actively seek solutions that foster a sustainable future. Social media allows them to bypass traditional media outlets and directly engage with the public, offering actionable steps for reducing carbon footprints and adopting eco-friendly practices.

This approach marks a significant departure from past generations' activism. Millennials are not simply content with raising awareness; they demand concrete solutions and advocate for systemic change. This can involve acts of civil disobedience, such as risking arrest to protest environmentally destructive practices – a tactic rarely employed by previous generations (Carrington, 2021). Social media is a powerful tool for documenting and disseminating these acts of defiance, inspiring others to join the fight for a sustainable future.

The landscape of climate activism has undergone a dramatic shift in recent years. While past generations expressed concern for the environment, millennials have emerged as a powerful force demanding action. This new wave of activism is characterized by a focus on solutions, a willingness to engage in civil disobedience, and the effective utilization of social media to spread awareness and mobilize support.

Several young, charismatic individuals are leading the charge and have captured global attention. Greta Thunberg, the now-iconic Swedish teenager, ignited a movement with her school strikes, urging world leaders to take immediate action on climate change. Her unwavering determination and powerful message resonated with millions, inspiring a generation to take up the mantle of environmental activism (Fridays for Future, 2023).

Thunberg is not alone. Jamie Margolin, founder of the youth-led organization Zero Hour, has mobilized young people across the United States to demand climate action (Zero Hour, n.d.). Varashini Prakash, co-founder of the Sunrise Movement, played a pivotal role in advocating for the Green New Deal, a comprehensive legislative package addressing climate change and economic inequality (Sunrise Movement, n.d.). These are just a few of the many young voices demanding change.

Beyond the United States, Indigenous activist and Earth Guardians Youth Director Xiye Bastida has used her platform to address the United Nations on environmental issues and advocate for including Indigenous voices in climate solutions (Lichtenstein, 2020). These young leaders represent a diverse group united by a common purpose – to ensure a livable planet for future generations.

Millennials' commitment to climate action extends beyond raising awareness. Frustrated with the lack of progress on traditional advocacy routes, they increasingly resort to civil disobedience to communicate their message. This willingness to risk arrest reflects the urgency of the climate crisis and their determination to see meaningful change.

A notable example is the ongoing campaign by Extinction Rebellion, a global network of activists employing nonviolent civil disobedience to compel governments to take action on climate change (Extinction Rebellion, n.d.). These tactics, while controversial, have garnered significant media attention and forced policymakers to confront the climate crisis head-on (Carrington, 2021).

Social media has become an indispensable tool for millennial activists. Platforms like Twitter, Instagram, and Facebook allow them to bypass traditional media outlets and connect directly with a global audience. They can share information about climate science, document acts of civil disobedience, and mobilize supporters for demonstrations and campaigns (Schroeder, 2020).

This digital activism fosters community and empowers individuals to act in their local contexts. Social media campaigns like #FridaysForFuture and #ClimateStrike have mobilized millions of people worldwide, demonstrating the collective power of this generation (Boykoff & Yulsman, 2013).

The rise of millennial climate activism represents a turning point in the fight for a sustainable future. Their innovative tactics, unwavering commitment, and effective use of technology propel the climate crisis to the forefront of global discourse. While challenges remain, the dedication and energy of this new generation of activists offer a beacon of hope in a time of unprecedented environmental crisis.

Millennials have emerged as a powerful force in the fight against climate change, not just through their activism but also through their strategic use of media and messaging. This generation understands the importance of capturing attention and sparking conversation – the cornerstones of any successful social movement.

By highlighting climate change issues, millennials are forcing them into the public sphere. This increased focus creates a platform for debate and discussion, which is crucial for enacting change. When the problem becomes a topic of conversation, it compels people to re-evaluate existing norms and question the status quo (Carrington, 2023). This pressure on those in power can ultimately lead to implementing new policies and regulations to address the climate crisis.

This phenomenon resonates deeply with me. While my children have been exposed to some climate change education in school, they haven't received in-depth instruction on the potential consequences of inaction. Their curriculum focuses on essential awareness rather than a comprehensive understanding of our challenges. This experience highlights the need for a more robust approach to climate change education in schools.

It is essential to equip younger generations with a thorough understanding of climate science and its implications. This knowledge empowers them to make informed choices, advocate for change, and become responsible stewards of the environment. A curriculum that goes beyond essential awareness and into potential solutions can inspire young minds to contribute to a more sustainable future (Sobel, 2020).

Education as a Catalyst for Change

Integrating climate change education into the core curriculum across all grade levels is crucial. This education should not be limited to rote memorization of scientific facts but rather foster critical thinking, problem-solving skills, and a sense of agency. By engaging students in discussions about climate solutions and encouraging them to explore potential career paths related to sustainability, we can create a generation equipped to tackle this complex challenge (Brewer, 2021).

Millennials have ignited a powerful movement, and education is the key to ensuring its long-term success. By empowering younger generations with knowledge and fostering a sense of responsibility, we can work together to build a more sustainable future for all.

My experience with my children's school project on recycling perfectly exemplifies the changing approach to environmental education. While recycling is a commendable practice, millennials recognize its limitations. The focus is shifting toward reducing plastic consumption altogether. This aligns with the concept of a circular economy, which prioritizes minimizing waste and maximizing resource use through product design, repair, and reuse (Ellen MacArthur Foundation, 2023).

Fortunately, this shift in perspective is being reflected in educational institutions. Environmental science programs are becoming increasingly common, equipping students with a deep understanding of climate change and its scientific underpinnings. This represents a stark contrast to the experiences of previous generations, who often received little to no exposure to these critical issues during their schooling.

The impact of this new curriculum extends beyond scientific literacy. Sustainability courses, for instance, explore the connection between dietary choices and climate change. This explains the phenomenon of young children advocating for plant-based diets – knowledge they likely acquired not from their parents but from their schools.

Formal education is just one piece of the puzzle. Millennials are fostering environmental consciousness through student-led initiatives. These initiatives take various forms, from campus clean-up projects to design competitions for innovative renewable energy technologies. This grassroots activism demonstrates a deep commitment to environmental responsibility and is a powerful driver of change within universities.

The tables are turning when it comes to climate knowledge. Millennials, raised in an era of heightened environmental awareness, are often more knowledgeable than their parents about climate issues. This has led to a fascinating dynamic where children educate their parents at the dinner table. The traditional flow of information is reversing, highlighting the effectiveness of millennial-driven climate education initiatives.

Millennials are not simply inheriting a climate crisis but actively shaping the conversation and driving solutions. Their focus on education, innovation, and activism fosters a generation deeply committed to environmental responsibility. This shift in perspective, evident in schools, homes, and universities, offers a glimmer of hope for a more sustainable future.

The Discussion Around Climate Change and Dietary Choices

The discussion around climate change has expanded to encompass dietary choices, with a particular focus on the environmental impact of meat consumption.

Environmental Impact of Beef

Methane Emissions: Cows belch methane, a greenhouse gas even more potent than carbon dioxide, when trapping atmospheric heat (Environmental Protection Agency, 2020).

Deforestation: Raising cattle necessitates vast tracts of land, often leading to deforestation of rainforests. Rainforests play a crucial role in absorbing carbon dioxide, and their destruction exacerbates climate change (Food and Agriculture Organization of the United Nations, 2023).

Shifting Consumer Preferences

Education: Millennials exposed to environmental science education make informed dietary choices based on sustainability concerns. Children are learning about the environmental impact of meat and advocating for plant-based options.

Nuances of Meat Consumption

Focus on Beef: The discussion should advocate for more than complete meat elimination but a reduction in beef consumption. Other protein sources like chicken, pork, or lamb are more sustainable when raised organically or in free-range.

Cost Considerations: Organic and free-range meat options are often more expensive, potentially burdening some consumers. However, proponents of plant-based diets argue that the long-term health benefits translate to reduced healthcare costs.

The Power of Social Media in Climate Change Advocacy

Social media platforms like Instagram, Twitter, and TikTok have emerged as powerful tools for spreading awareness about climate change and mobilizing action. Millennials, adept at navigating these digital spaces, are leveraging them to:

Disseminate Information: Climate science research, environmental destruction visuals, and calls to action are readily shared across vast networks, fostering global awareness of the crisis.

Spark Conversation: Social media platforms create spaces for discussion and debate, encouraging critical thinking and engagement with climate issues.

Organize Movements: Millennials leverage social media to organize demonstrations, promote petitions, and coordinate advocacy efforts, amplifying their collective voice (Boykoff & Yulsman, 2013).

Even if immediate action proves difficult, the shift in consciousness is a crucial first step. By recognizing the gravity of the situation, individuals are more likely to embrace sustainable practices in the future. This awareness can also fuel innovation and motivate policymakers to prioritize climate solutions.

The story of MrBeast, a young YouTuber who used his resources to drill wells for clean water access in Africa, exemplifies the power of individual action (MrBeast Philanthropy, n.d.). His initiative highlights the importance of empowering individuals to become agents of change. Moving beyond simply sending aid, Mr. Beast's project equips villages with the tools to become self-sufficient and improve their long-term well-being. This approach aligns with the proverb, "Give a man a fish, and you feed him for a day; teach a man to fish, and you feed him for a lifetime" (Knuth, 2008).

Social Media Strategies for Maximizing Your Climate Impact

Social media platforms offer individuals a powerful tool for advocating for climate action. By following these best practices, millennials can leverage their online presence to spark meaningful conversations and inspire concrete steps within their communities.

Importance of Climate Advocacy

The Urgency of the Climate Crisis: The scientific consensus overwhelmingly affirms the urgency of addressing climate change. Individual action, scaled up through social media engagement, can contribute significantly to collective efforts (Intergovernmental Panel on Climate Change, 2021).

Crafting Effective Social Media Posts

Evidence-Based Approach: Backing your message with credible sources, such as scientific reports or reputable news articles, strengthens its impact (Boykoff & Yulsman, 2013).

Conciseness is Key: Short, attention-grabbing posts are more likely to be read and shared on social media platforms (Lennox & Sacks, 2014).

Actionable Message: End your posts with a clear call to action, encouraging your followers to take a specific step, like signing a petition or adopting a sustainable practice (McCombs & Shaw, 1972).

Engaging Your Online Community

Be a Spokesperson, Not an Activist: While traditional activism tactics like protests are valuable, social media advocacy allows individuals to champion climate action from their own spaces (Schroeder, 2020).

Share Relevant Content: Curating and sharing posts from reputable climate organizations and thought leaders keep your followers informed and engaged.

Call Out Unsustainable Practices: Social media lets you gently point out environmentally harmful behavior and encourage positive change within your network.

Be Vulnerable and Authentic: Sharing your journey of environmental awareness can inspire others to act.

By following these strategies, millennials can harness the power of social media to promote climate action and foster a culture of environmental responsibility within their online communities. Remember, even small actions, amplified through social media, can make a significant difference in the fight against climate change.

The world needs your voice. Climate change, social justice, animal welfare – our challenges are vast, but so is your potential to make a difference. The power of advocacy lies within you, waiting to be unleashed. However, where do you begin?

Finding Your Voice

Imagine you have the opportunity to give a short speech about an environmental issue you care about for your entire school. What would you say? How would you frame your message to be informative, engaging, and inspiring to your peers? Write down your speech, focusing on the specific actions you would encourage your classmates to take.

Social Media Savvy

Social media platforms can be powerful tools for advocacy. Think about a specific environmental issue you are passionate about. Design a social media campaign (choose the platform) to raise awareness. Sketch some sample posts, considering visuals, hashtags, and a call to action.

Leading by Example

Reflect on your daily habits. Identify areas where you could reduce your environmental impact. Brainstorm a plan to implement these changes, considering how you can overcome any challenges. How can you inspire others to adopt more sustainable practices?

Knowledge is Power

A strong advocate requires a solid understanding of the issue. Choose an environmental topic you would like to learn more about—research credible sources of information, such as scientific reports, documentaries by reputable organizations, etc. Summarize your key takeaways and brainstorm how this new knowledge could inform your advocacy efforts.

Collaboration is Key

Change often happens most effectively through collective action. Think about environmental organizations or initiatives in your community. Research their work and consider how your skills and interests could contribute to their efforts. Would you prefer to volunteer, participate in events, or spread awareness through your network?

In conclusion, Millennials, armed with the power of technology and social media, have ignited a global conversation about climate change. Their innovative use of digital platforms has fostered unprecedented awareness and inspired action across generations. However, one generation cannot win the fight for a sustainable future alone.

Bridging the generational divide is crucial. Millennials' tech-savvy approach, coupled with the wisdom and experience of older generations, presents a formidable force for change. Through open communication and a shared understanding of the urgency of the climate crisis, we can create a united front for a healthier planet.

This journey does not require grand gestures. As the next chapter explores, even small steps, when taken collectively, can lead to significant change. Together, let us leverage the power of collaboration and embrace the responsibility to create a livable future for future generations.

CHAPTER 5
Bridging Climate Divides: Perspectives From First And Third World Nations

The world faces a significant challenge in addressing climate change. While the urgency of the crisis is undeniable, a one-size-fits-all approach fails to consider the vast differences between developed and developing countries. This chapter explores this critical disparity, highlighting the unique challenges developed nations face and the crucial role developed nations must play in fostering a collaborative path toward a sustainable future.

A quote by Ban Ki-moon, former Secretary-General of the United Nations, perfectly encapsulates the spirit of this chapter: "If we want to tackle climate change, we need to work together. No one country can do it alone." Developed nations possess a wealth of resources, technological advancements, and knowledge that can empower developing countries to adapt to climate change and build resilience.

By bridging the gap through financial aid, technology transfer, and knowledge sharing, we can create a global agricultural sector that is productive and sustainable for generations to come.

Let us consider the daily struggles faced by individuals in developing countries. Their primary concern is securing necessities – putting food on the table – leaving little bandwidth for anxieties about climate change. A 2021 study published in Nature Climate Change underlines this point. Researchers found that "individuals living in poverty prioritize immediate threats to their livelihoods and well-being over longer-term environmental concerns" (Carrico et al., 2021).

This does not imply a lack of concern for the environment. However, the urgency of survival takes precedence. This is where developed nations can play a crucial role. We can offer practical suggestions for reducing carbon footprints without jeopardizing their immediate needs.

The disparity extends beyond daily life. Developed nations hold a significant advantage in technological advancements. Renewable energy solutions, smart grids, and carbon capture systems still need to be available to developing countries. A 2023 report by the International Renewable Energy Agency (IRENA) highlights this gap, stating that "developing countries require an annual investment of USD 1.3 trillion by 2030 to achieve their renewable energy targets and contribute to limiting global warming to 1.5°C" (IRENA, 2023).

This financial reality creates a vicious cycle. Developing countries need more capital to invest in research and development for cutting-edge technologies that could propel them toward a more sustainable future. The burden of innovation falls heavily on developed nations, who then hold the intellectual property rights to these advancements.

Policy and regulations further exacerbate the gap. Developed nations often have stringent environmental regulations and comprehensive climate policies. These frameworks allow them to impose stricter emission targets, incentivize green technology adoption, and participate in international agreements like those under the European Union. Developing countries, needing more resources to enforce such measures, need help to gain traction in their fight against climate change.

Public awareness also plays a critical role. In developed nations, extensive campaigns and educational initiatives have fostered a robust societal understanding of climate change. This awareness translates into informed decision-making and a collective push for sustainable practices.

However, a 2022 study by the World Bank underscores the educational gap. The report found that "climate change education remains limited in many developing countries, particularly in rural areas" (World Bank, 2022). This lack of awareness hinders their ability to prioritize climate action and implement effective solutions.

The consequences of this disparity are stark. Developing countries often bear the brunt of climate change's immediate effects. While developed nations receive advanced weather forecasting, allowing them to prepare for storms and floods, developing countries are left scrambling to react to these disasters. The lack of resources for early warning systems and disaster preparedness leaves them incredibly vulnerable.

Economic constraints further complicate the situation. Developing nations typically need more financial resources, making investing in expensive climate mitigation technologies nearly impossible. Developed countries, on the other hand, can leverage government grants and subsidies to support sustainable practices within their agricultural sectors and industries. A 2023 study published in Environmental Science & Policy found that "increased financial aid from developed countries to developing countries is crucial for accelerating the transition to a low-carbon economy" (Rogelj et al., 2023).

The path forward demands a multifaceted approach. Developed nations must go beyond simply offering aid. Technology transfer, where knowledge and expertise are shared with developing countries, is essential. This fosters long-term sustainability by empowering these nations to build solutions and transition toward a greener future.

Imagine a rural community in Southeast Asia. The villagers were incredibly aware of their environmental challenges, such as deforestation and soil erosion. However, their primary concern was securing enough food to feed their families. The project provided them with training on sustainable farming practices that improved their yields without further burdening the environment. This experience highlighted the importance of tailored solutions that address the immediate needs of developing communities while laying the groundwork for a more sustainable future.

By acknowledging the vast differences in priorities and resources between developed and developing countries, we can craft a more compelling narrative around climate action. Developed nations must be responsible for technological innovation, financial assistance, and knowledge transfer. We can only achieve a sustainable future for our planet through a collaborative effort that recognizes the unique challenges faced by each nation.

Financial Aid, Technological Solutions, and Food Security

Developed nations have a crucial role in supporting developing countries as they grapple with the challenges of climate change. While some efforts exist, the current approach must be revised to meet the needs.

One area ripe for improvement is intellectual property rights. Developed nations can explore easing patent laws on climate-mitigating technologies. This would allow developing countries to access and adapt these advancements to their needs. Subsidies can also play a vital role. Just as some governments, like China, offer financial support to their agricultural communities, developed nations can provide similar subsidies to incentivize sustainable practices in developing countries.

Financial support extends beyond subsidies. Developed nations can contribute significantly by increasing their contributions to climate funds. These funds can then be used to invest in critical sustainable infrastructure projects in developing countries.

Examples include dams for improved water management, river clean-up initiatives to combat pollution, and capacity-building programs to equip local communities with the knowledge and skills to adapt to a changing climate. Knowledge sharing is another crucial aspect of this support. Developed nations can foster international collaboration by sharing scientific advancements, best practices, and educational resources with their developing counterparts.

Food security is another critical area demanding international collaboration. Climate change disrupts food production patterns, impacting imports, exports, and global food availability. Developed nations can play a vital role in mitigating these disruptions. One approach involves supporting the development and dissemination of genetically modified organisms (GMOs). These crops, engineered to thrive in harsher climates, offer a potential solution for ensuring food security in vulnerable regions.

Beyond GMOs, international collaboration can bridge the gap in agricultural techniques. Developed nations can share expertise and resources to introduce drought-resistant crop varieties in regions prone to water scarcity. This knowledge transfer and sharing of sustainable farming practices can significantly enhance food security, freshwater availability, and resilience in developing countries.

Imagine the devastating impact of climate change on a small farming community in Africa. Erratic rainfall patterns have led to severe droughts, jeopardizing crop yields and threatening food security. However, a ray of hope emerged through an international collaboration. An agricultural research institute from a developed nation partnered with the local community to introduce drought-resistant maize varieties. This intervention proved successful, ensuring a more stable food supply for the community. This experience underscored the power of international collaboration in fostering resilience and promoting sustainable practices in the face of climate change.

By taking a multifaceted approach – easing patent restrictions, offering financial support, sharing knowledge, and collaborating on agricultural advancements – developed nations can empower developing countries to adapt to climate change and build a more sustainable future. The current efforts represent a starting point, but a significant increase in financial and technological support is necessary to bridge the gap and ensure a future where food security and environmental well-being are attainable for all nations.

Climate change significantly threatens water security worldwide, particularly in developing nations. Developed countries have a crucial role in helping these nations build resilience through investments in water infrastructure, economic support for transitioning to sustainable practices, and disseminating effective water management technologies.

One critical area for investment is water infrastructure. Building dams, implementing efficient water management systems, and providing the necessary pumping technology are expensive endeavors. Developed nations can provide financial and technical assistance to developing countries to undertake these crucial projects. The benefits are undeniable, as these improvements ensure access to clean water for domestic and agricultural purposes.

The economic impacts of climate change are another area demanding attention. Extreme weather events can weaken entire economies, as evidenced by the devastation caused by the recent hurricane in Puerto Rico. Developed nations can be vital in supporting economic recovery after such disasters. This includes providing financial aid, supporting infrastructure reconstruction, and collaborating on long-term climate-resilient development plans.

Transitioning to sustainable practices is crucial for long-term economic stability in the face of climate change. Developed nations can provide financial incentives and technical expertise to support developing countries adopting sustainable agricultural practices and renewable energy sources. This reduces vulnerability to climate shocks and fosters economic growth in a low-carbon future.

Beyond large-scale infrastructure projects, numerous cost-effective solutions can make a significant difference. Rainwater harvesting is a prime example. Simple rooftop collection systems and storage tanks can provide a reliable water source during dry seasons. This approach is practical and financially accessible for developing countries and individual communities.

My father, recognizing the changing climate patterns with longer and hotter summers, built a rainwater catchment system on our property over 30 years ago. This system collects rainwater from the roofs of our house and farm buildings, storing it in a large underground tank with filters. The collected water is pumped into the house for use in toilets, washing machines, and kitchen plumbing. This system has proven remarkably effective and cost-efficient.

This personal experience highlights the power of simple solutions. Rainwater harvesting offers numerous benefits. It is a readily available resource, requires minimal maintenance, and reduces dependence on expensive municipal water supplies. Rainwater is also naturally soft, containing beneficial minerals and requiring less soap for laundry, further enhancing its appeal. Unfortunately, my father's experience also highlights potential challenges. Despite its inherent sustainability, local authorities may attempt to regulate or tax rainwater collection.

Microfinance, Farmer Perspectives, and Sustainable Practices

Microfinance offers a promising approach to building climate resilience in developing countries. This system provides small financial grants to villages and communities, which can fund various climate-adaptation projects, such as drilling wells, installing water pumps, building irrigation ditches, and purchasing livestock feed for farmers. By empowering local communities to address their specific needs, microfinance enables them to implement solutions that enhance their ability to withstand the challenges posed by climate change (Moyo, 2006).

To gain a deeper understanding of the impact of climate change on agriculture, let us explore the contrasting perspectives of farmers in developed and developing nations. A Chinese farmer in 2024 faces a blend of immediate concerns, including environmental changes like erratic rainfall patterns that threaten crop yields, pests and diseases that become more prevalent due to a disrupted ecosystem, and economic pressures that mount due to fluctuating market prices and import/export issues. Government policies and support for infrastructure development, education, and sustainable practices also shape this farmer's daily realities (Li et al., 2021).

In contrast, a British or American farmer may focus more on optimizing efficiency and profitability. While concerns about climate change are likely present, they may be overshadowed by technological advancements, labor costs, and market competition. These farmers typically have access to advanced technologies, government subsidies, and educational resources that can help them adapt to a changing climate (European Commission, 2023; United States Department of Agriculture, 2023).

Economic disparity undoubtedly plays a role in these contrasting perspectives, but awareness is also crucial. Some farmers in developing countries may need more access to resources or financial means to implement sustainable practices. In contrast, others may need to be made aware of the environmental consequences of their actions. For example, burning crop residue, a common practice in regions like Punjab, India, and parts of China and Pakistan, contributes significantly to air pollution (Guttikunda et al., 2023). Although this practice may seem quick and inexpensive to clear fields for the next planting season, its long-term effects can be devastating.

This highlights the importance of education and awareness campaigns. Equipping farmers with the long-term consequences of their actions is crucial for promoting sustainable practices. Developed nations can contribute to knowledge transfer by sharing best practices, educational resources, and technological advancements with farmers in developing countries.

Adapting to Change

Climate change presents a significant financial risk to farmers around the world. As weather patterns become increasingly unpredictable, crop yields suffer, and the threat of disease outbreaks intensifies. Livestock farmers face similar challenges, with the potential for entire herds to be wiped out by disease due to changing environmental conditions. These risks are further compounded by rising insurance premiums, placing an additional financial burden on farmers who are already struggling (Azeez, 2023).

One potential solution lies in carbon trading and incentive programs. The UK government's initiative to encourage carbon sequestration practices in agriculture is a prime example. By offering carbon credits to farmers who engage in activities like planting hedgerows, restoring peatlands, or clearing waterways, the government provides a financial incentive for them to adopt sustainable practices. These carbon credits can offset emissions elsewhere, creating a win-win situation for the environment and farmers' incomes (UK Department for Environment, Food & Rural Affairs, 2023).

Technology is another powerful tool in the fight against climate change. British farmers increasingly adopt technologies like satellite imagery and data analytics to optimize resource use and diversify their yields. This approach can help them adapt to changing weather patterns and identify areas where their crops are most likely to thrive. Additionally, some farmers are exploring the potential of raising new livestock breeds better suited to the changing climate (Gilmour et al., 2021).

Diversification extends beyond crops and livestock. Agri-tourism, which involves offering tours of farms or renewable energy projects, is gaining traction as a way to generate additional income. This approach benefits farmers financially and fosters public awareness about the challenges and opportunities presented by climate change (Getz & Pickering, 2023).

Building resilience also requires strong social and community collaboration. Farmer cooperatives and barter exchange programs allow farmers to share resources, buy and sell goods directly from each other, and reduce their reliance on traditional market channels. This collaborative approach can significantly reduce transportation costs and ensure excellent market stability for farmers (Iles, 2020). Sharing best practices and knowledge allows farmers to learn from each other's experiences and adapt more effectively to a changing climate.

The examples from British agriculture offer valuable insights for farmers around the world. By embracing innovative financial incentives, technological advancements, and collaborative practices, farmers can build resilience, mitigate climate risks, and ensure the long-term sustainability of their livelihoods.

The innovative practices adopted by British farmers offer valuable lessons for developing countries grappling with climate change challenges. Closing the gap between these nations requires a multifaceted approach that includes knowledge sharing, financial investment, and technology transfer.

Developed nations can play a crucial role in disseminating knowledge and best practices. The information covered in this chapter, including financial incentives like carbon trading programs, the use of technology for resource optimization, and the power of collaborative farming practices, can all be valuable tools for farmers in developing countries. Educational programs, workshops, and agricultural extension services can equip these farmers with the knowledge and skills to adapt to climate change and build resilience (FAO, 2023).

Financial investment from the private sector can be another powerful driver of change. While philanthropic efforts are commendable, a compelling business case exists for investing in developing-world agriculture. These economies are often heavily reliant on agriculture, and by investing in new technologies and sustainable practices, private entities can generate healthy profits and contribute to a more stable and climate-resilient global food system (The World Bank, 2023).

The role of influential figures like Elon Musk and Bill Gates is also noteworthy. These individuals have the resources and public profiles to champion sustainable agriculture in developing countries. By investing in this sector, they can generate positive financial returns and enhance their reputations as responsible global citizens. This approach can attract other investors and encourage a broader shift toward sustainable agricultural practices.

Additional Considerations

Focus on building local capacity: Investment and technology transfer should not create a dependency on external resources. The goal should empower developing countries to develop their expertise and infrastructure in sustainable agriculture.

Equity and inclusion: Investment and knowledge sharing should reach smallholder farmers, who are often the most vulnerable to climate change but also play a critical role in food security.

Policy and regulatory frameworks: Developing countries may need to adjust their policies and regulations to incentivize sustainable practices and create a more attractive environment for private sector investment.

By working together, developed nations, the private sector, and influential individuals can bridge the gap between developed and developing countries in the fight against climate change. By promoting knowledge sharing, financial investment, and technology transfer, we can foster a global agricultural sector that is productive, profitable, sustainable, and resilient in the face of climate challenges.

1. Considering the Needs of Different Communities:

Imagine you are an agricultural extension agent working with a community of farmers in a developing country. They rely heavily on traditional farming methods, but climate change disrupts their yields. Research a specific sustainable practice - like rainwater harvesting or cover crops - that could benefit this community.

a) Considering their priorities and resource limitations, how would you explain the benefits of this practice to the farmers?

b) What challenges might they face in adopting this practice?

c) How could you address those challenges and encourage them to give it a try?

2. Financial Responsibility and Sustainable Choices:

The text mentions carbon trading programs as a financial incentive for farmers to adopt sustainable practices. Think about your buying habits.

a) Do you consider the environmental impact of the products you purchase?

b) Are there ways you could adjust your shopping habits to support sustainable agriculture?

c) How could you encourage others to make similar choices?

3. Building Resilience Through Collaboration:

The text discusses the importance of collaboration among farmers. Reflect on your community or social circles.

a) Are there ways you could collaborate with others to reduce your environmental impact or become more sustainable?

b) This could involve sharing resources with neighbors to start a community garden.

c) Brainstorm some ideas and consider how you could get others involved.

Climate Control Projects Around the World

Climate change is a pressing issue facing our planet. Countries and organizations worldwide are developing various creative and practical solutions. This section will explore some of these remarkable climate control projects, highlighting the diversity and ingenuity applied to this global challenge.

Scientists and organizations around the world are tackling climate change from multiple angles. Here, we explore a few examples that showcase the diversity and ingenuity of these efforts:

A. Solar Radiation Management Research

This approach investigates methods to reflect a small portion of incoming sunlight into space, potentially mitigating global warming. International research teams are exploring various techniques. For instance, the Stratospheric Particle Injection for Climate Engineering (SPICE) project in the UK investigates injecting aerosols into the stratosphere to achieve this effect. Similarly, Harvard University's Solar Geoengineering Research Program explores solar radiation management technologies' feasibility and potential risks (*Harvard's Solar Geoengineering Research Program*, n.d.).

B. The Great Green Wall, Africa

This ambitious pan-African initiative aims to grow an 8,000-kilometer wall of trees across the entire width of the African continent (*Growing a World Wonder - Great Green Wall*, n.d.). The project seeks to combat desertification, a process that degrades fertile land into desert, by increasing vegetation cover. This improves food security for local communities and helps them adapt to the changing climate.

C. The Ocean Cleanup, International Waters

Plastic pollution in our oceans poses a significant threat to marine ecosystems. The Ocean Cleanup, a non-profit organization headquartered in the Netherlands, tackles this issue by developing and deploying advanced technology to collect plastic debris from the world's oceans (The Ocean Cleanup, 2024). One of their central projects targets the Great Pacific Garbage Patch, a vast accumulation of plastic waste in the North Pacific Ocean.

D. Smart Forest City, Cancun, Mexico

This proposed development in Cancun, Mexico, aims to establish a model for sustainable urban living (Stefano Boeri Architetti, 2023). The city would integrate nature into its urban planning, featuring advanced water recycling systems and a reliance on solar energy and other renewable resources. Such projects demonstrate the possibility of creating thriving urban centers with minimal environmental impact.

E. Tesla's Global Impact

Tesla, a company known for its electric vehicles (EVs), is a significant player in accelerating the transition to sustainable energy. According to Tesla's Impact Report 2023, their electric cars and battery storage systems offer a considerable advantage over traditional gasoline-powered vehicles by:

Reducing carbon emissions: Electric cars produce zero tailpipe emissions, directly contributing to cleaner air and reduced greenhouse gas emissions.

Promoting renewable energy: Tesla's battery storage systems allow for integrating renewable energy sources like solar and wind power into the grid, making renewable energy more reliable and efficient.

Tesla's global presence and focus on innovation drive the electric vehicle market forward, paving the way for a more sustainable transportation future by demonstrating their feasibility, environmental benefits, and increasing accessibility for consumers worldwide.

F. Energiewende, Germany

Germany's Energiewende, or "energy transition," is a comprehensive policy aiming to shift the country's energy production from fossil fuels to renewable sources. This ambitious initiative includes:

Phasing out nuclear and coal power: Germany plans to eliminate nuclear power by 2022 and phase out coal-fired power plants by 2038.

Investing heavily in renewable energy: Germany is a global leader in solar and wind power generation, with ambitious targets to increase their share of the energy mix.

Promoting energy efficiency: Policies and incentives encourage citizens and industries to adopt energy-saving technologies and practices.

The Energiewende is a model for other countries seeking a significant shift toward renewable energy sources.

G. Forest City, Malaysia

Forest City, located in Malaysia, is a large-scale development project designed to be a sustainable urban center. The project incorporates various features to minimize environmental impact and create a "living green lung":

Green building practices: Buildings are constructed with sustainable materials and designed for energy efficiency.

Waste management systems: Advanced technologies minimize landfill waste and promote recycling.

Urban forestry: Integrating parks, green spaces, and vertical gardens throughout the city improves air quality and provides natural habitats.

Forest City demonstrates how urban development can be designed with environmental sustainability.

H. Masdar City, UAE

Masdar City, located in the United Arab Emirates (UAE), is an ambitious project aiming to be the world's first zero-carbon city. The development focuses on clean technology and sustainable practices:

Clean technology hub: Masdar City attracts businesses and research institutions focused on clean energy, water conservation, and sustainable materials.

Traditional architecture with modern technology: Buildings incorporate elements of traditional Arabian architecture that promote natural ventilation and cooling, combined with modern technologies for energy efficiency.

Renewable energy sources: The city relies heavily on solar power and other renewable energy sources to minimize its carbon footprint.

Masdar City serves as a blueprint for sustainable urban development in arid regions.

I. Reforestation Projects in the Amazon, South America

The Amazon rainforest plays a crucial role in global climate regulation. Several initiatives are underway to address deforestation and promote reforestation in the Amazon basin:

Restoration projects aim to replant trees in degraded rainforest areas, helping restore biodiversity, improve soil health, and recapture atmospheric carbon dioxide.

Indigenous community involvement: Empowering Indigenous communities who have traditionally stewarded the rainforest is critical to successful reforestation efforts.

Sustainable development programs: Programs that provide economic alternatives to deforestation, such as sustainable forestry practices and non-timber forest product harvesting, are crucial for long-term success.

Reforestation efforts in the Amazon are essential for mitigating climate change and preserving the ecological health of this vital biome.

J. Iceland's Carbfix Project

Iceland's Carbfix Project takes a novel approach to addressing climate change by capturing and storing carbon dioxide (CO_2) emissions underground:

Capturing CO_2 emissions: Carbfix captures CO_2 emissions from industrial processes, such as geothermal power plants.

Mineralization: The captured CO_2 is dissolved in water and injected deep underground, reacting with rock formations to permanently store the carbon as a stable mineral.

Safe and long-term storage: This technology offers a promising solution for permanently removing CO_2 from the atmosphere and mitigating climate change.

The Carbfix Project demonstrates the potential of carbon capture and storage technologies for combating climate change.

These projects showcase the diversity of approaches undertaken worldwide to address climate change. The following section will explore how developed and developing countries implement these solutions.

Examples by Development Status

The fight against climate change requires a global effort, and countries at various development stages are implementing different solutions. Here is a look at how some approaches are being applied:

A. Solar Power Projects

Solar energy offers a clean and abundant renewable energy source. Both developed and developing countries are making significant strides in solar power generation:

Developed Countries: Germany and the Netherlands are global leaders in solar energy adoption, with extensive solar infrastructure and ambitious targets for increasing their share of renewable energy (International Renewable Energy Agency, 2023).

Developing Countries: India's ambitious solar power initiative aims to generate 40% of its energy from non-fossil fuel sources by 2030, with solar playing a pivotal role (MNRE, Government of India, 2023).

B. Sustainable Transportation

Sustainable transportation solutions are crucial for reducing greenhouse gas emissions. Here are some examples:

Developed Countries: Cities like Copenhagen, Denmark, and Amsterdam, Netherlands, have prioritized cycling infrastructure and efficient public transportation systems, encouraging cleaner commutes (Marchetti et al., 2017).

Developing Countries: Bogota, Colombia, implemented a successful Bus Rapid Transit (BRT) system, offering a reliable and affordable alternative to private car use (Institute for Transportation and Development Policy, 2013).

C. Waste Management and Recycling

Efficient waste management and recycling are crucial for mitigating environmental impact. Here are some examples:

Developed Countries: Sweden is a leader in waste-to-energy technology, converting waste into heat and electricity and reducing reliance on fossil fuels (Eurostat, 2022). Japan has implemented highly efficient recycling systems to minimize waste generation (JMETI, 2023).

Developing Countries: While many developing countries grapple with waste management challenges, some initiatives are emerging. For example, Costa Rica's waste management program has implemented successful recycling and composting initiatives.

D. Reforestation and Forest Conservation

Forests play a vital role in carbon sequestration and maintaining ecological health. Both developed and developing countries are engaged in reforestation and conservation efforts:

Developed Countries: Canada and Norway have vast forest resources and are implementing policies to protect and sustainably manage them (Natural Resources Canada, 2023; Norwegian Environment Agency, 2023).

Developing Countries: Costa Rica's Payment for Environmental Services (PES) program financially rewards landowners for conserving and restoring forests, providing a successful model for developing countries (Engel et al., 2008).

Conclusion

The fight against climate change requires a global effort, and the projects highlighted here showcase the diversity and ingenuity being applied worldwide. Developed and developing countries are implementing innovative solutions to reduce greenhouse gas emissions, adapt to a changing climate, and protect our planet.

International collaboration is crucial for tackling this complex challenge. Sharing knowledge, resources, and best practices can accelerate progress toward a sustainable future. Many more inspiring climate control projects exist beyond the scope of this section. From community-driven renewable energy initiatives to technological advancements in carbon capture, the global landscape of climate action is constantly evolving.

The emphasis must be on continued international cooperation and a shared commitment to sustainable practices. By working together, countries can create a cleaner, healthier planet for future generations.

CHAPTER 6
What Of All The People: The Flood Of Climate Refugees

Climate change disrupts lives and communities in unprecedented ways. One of the most concerning consequences is the displacement of people from their homes due to environmental factors.

However, defining "climate refugee" is a complex issue. The term is not currently recognized in international law, and displacement often occurs due to a combination of factors, making it difficult to isolate climate change as the sole cause.

Who Are Climate Refugees?

Despite the definitional challenges, a general understanding of climate refugees exists. The United Nations Environment Programme (UNEP) defines them as those forced to leave their homes "because of marked environmental disruption" (El-Hinnawi, 1985).

This disruption can take many forms, including rising sea levels, extreme weather events, desertification, and water scarcity. These environmental changes threaten livelihoods, food security, and safety, pushing people to migrate for a better life.

The Science Behind the Movement

Several key factors, including rising sea levels, increased frequency and severity of weather events, and desertification, drive the environmental changes that lead to displacement. Here is an explanation of how each of these factors contributes to displacement:

Rising Sea Levels: A primary consequence of climate change is the thermal expansion of oceans and melting glaciers and ice sheets. This translates to rising sea levels, which threaten coastal communities and low-lying island nations. Saline water intrusion contaminates freshwater sources and agricultural land, while inundation disrupts infrastructure and destroys homes. A study by Nicholls et al. (2011) predicts that by 2100, sea levels could rise by up to 1 meter, displacing millions residing in coastal areas.

Increased Frequency and Severity of Weather Events: Climate change disrupts weather patterns, leading to more frequent and intense storms, floods, droughts, and heat waves. These extreme weather events wreak havoc on infrastructure, cause widespread property damage, and disrupt agricultural production.

Desertification: Rising temperatures, coupled with changes in precipitation patterns, are accelerating desertification, the process by which fertile land degrades into arid desert. This phenomenon reduces agricultural productivity, destroys grazing land, and limits access to fresh water. The United Nations Convention to Combat Desertification (UNCCD, 2021) estimates that desertification directly affects over 3.6 billion people globally, with millions at risk of displacement due to land degradation.

Case Studies: The Human Cost of Environmental Change

Small Island Nations: Island nations like Kiribati and Tuvalu are at the forefront of the climate crisis. Rising sea levels threaten to engulf entire islands, displacing populations and submerging cultural heritage. These nations face an existential threat, with limited options for relocation and adaptation (McNamara & Connell, 2021).

Coastal Cities: Densely populated coastal cities are particularly vulnerable to rising sea levels and storm surges. Cities like Karachi, Mumbai, Jakarta, and Miami face the risk of massive displacement due to inundation and coastal erosion. Climate change also exacerbates inequalities within these cities, as low-income communities often reside in flood-prone areas (Satterthwaite, 2018).

Arid and Semi-Arid Areas: Regions like the Sahel in Africa and the Middle East are experiencing increased desertification and water scarcity. This limits agricultural productivity and strains water resources, pushing communities toward migration for sustainable livelihoods (UNCCD, 2021).

What Do the Numbers Tell Us?

Quantifying the exact number of climate refugees remains challenging due to the complex interplay of factors driving migration. However, estimates paint a concerning picture. The Internal Displacement Monitoring Centre (IDMC) reports that in 2022 alone, weather-related disasters displaced over 40 million people internally within their countries (IDMC, 2023).

The World Bank estimates that 2050 climate change could displace 216 million people within their countries (World Bank, 2023). These numbers highlight the escalating crisis and the urgent need for solutions.

Beyond the Numbers:

Statistics offer a broad picture, but the true impact of climate displacement is best understood through personal stories.

Imagine Amina, a Mali farmer who was forced to leave her ancestral land due to worsening droughts and desertification. Her once-fertile fields have become barren, leaving her family with no choice but to search for a new life elsewhere (The New Humanitarian, 2023).

Alternatively, consider Miguel, a fisherman in the Philippines whose coastal village was devastated by a powerful typhoon fueled by rising sea levels. He lost his home and livelihood, becoming an unwelcome addition to overcrowded urban centers (Deutsche Welle, 2022). These are just two examples of the countless lives shattered by climate change.

What Are the Socio-Economic Impacts of Displacement?

The mass displacement of people triggered by climate change has significant socio-economic consequences for both displaced populations and host communities. Displaced individuals often arrive in new locations with limited resources and skills, struggling to find adequate housing, food, and employment. This can increase resource competition in host communities, straining social services and infrastructure.

A World Food Programme (WFP, 2023) study warns that climate displacement exacerbates food insecurity, particularly in developing countries. Disruptions to agricultural production due to extreme weather events further complicate the situation. The mental health toll of displacement is also significant, with refugees experiencing trauma, loss, and uncertainty about their future.

What Are the International Responses to Climate Change Displacement?

Despite the growing number of climate refugees, there is currently no comprehensive international legal framework specifically designed to protect them.

The 1951 Refugee Convention is the cornerstone of international refugee protection. It defines a refugee as someone who has fled their country due to a "well-founded fear of persecution based on race, religion, nationality, membership of a particular social group or political opinion" (United Nations High Commissioner for Refugees [UNHCR], 1951).

However, this definition does not explicitly encompass displacement caused by environmental factors. While some argue that persecution due to ecological degradation could fall under the existing framework, the legal interpretation remains ambiguous and contested (Betts, 2020).

This lack of a clear legal definition creates a protection gap for climate refugees. They often struggle to access fundamental rights and humanitarian assistance available to refugees under international law. Furthermore, the current system primarily focuses on displacement across national borders, neglecting the significant internal displacement within a country's territory due to climate change (Warner, 2014).

Exploring Innovative Local Solutions

The limitations of international law have spurred the development of innovative local and community-driven solutions to address climate displacement. These solutions often focus on building resilience and adaptation strategies in vulnerable communities.

Bangladesh: This nation, highly susceptible to rising sea levels and cyclones, has implemented a community-based approach to climate change adaptation. The "polder system" involves raising land through embankments to protect villages from flooding. This initiative empowers local communities to manage their environment and reduce displacement risks (World Bank, 2020).

Pacific Island Nations: With existential threats from rising sea levels, Pacific Island nations are exploring innovative solutions like land buyouts and managed retreat programs. These programs aim to assist communities in relocating to higher ground and rebuilding their lives in safer areas (The Guardian, 2021).

Kenya: Recognizing the impact of climate change on pastoralist communities, Kenya has adopted a mobile phone-based early warning system. This system alerts herders about impending droughts and floods, allowing them to plan their movements and protect their livestock, a crucial source of income and food security (International Institute for Sustainable Development [IISD], 2023).

What Are the Projections of Future Displacement?

Climate scientists predict that the frequency and intensity of extreme weather events will increase as global temperatures continue to rise. This will exacerbate existing vulnerabilities and trigger large-scale displacement.

As mentioned, the World Bank estimates that under a business-as-usual scenario, with no significant mitigation efforts, climate change could force up to 216 million people to migrate internally within their own countries by 2050. Other estimates range even higher, with the International Institute for Environment and Development (IIED, 2023) predicting a staggering one billion climate refugees by 2050.

These numbers highlight the potential scale of the crisis we face and underscore the urgency of taking immediate action to mitigate climate change and develop strategies to manage displacement effectively.

Where Will the Crisis Be Most Acute?

Certain regions are particularly vulnerable to climate change's displacement impacts. These areas often face environmental threats, including rising sea levels, extreme weather events, desertification, and water scarcity.

Low-Lying Island Nations: Island nations like Kiribati and the Maldives face existential threats from rising sea levels. Entire islands could become submerged, displacing entire populations and wiping out cultural heritage (McNamara & Connell, 2021).

Densely Populated Coastal Cities: Coastal megacities like Mumbai, Lagos, and Shanghai are at risk from inundation and storm surges. These cities often lack adequate infrastructure to manage rising sea levels and extreme weather events, potentially leading to mass displacement of millions of residents (Satterthwaite, 2018).

Fragile States and Conflict Zones: Climate change exacerbates vulnerabilities in fragile states and conflict zones. Droughts and food insecurity can fuel tensions and social unrest, further disrupting livelihoods and displacing populations struggling with instability (Kelley, Oppenheimer, & Pettorelli, 2015).

The Devastating Impact of a Two °C Temperature Rise on Displacement

A global temperature rise of just 2 degrees Celsius might seem insignificant. However, the consequences for human populations, particularly regarding displacement and the creation of climate refugees, will be profound. This section details the estimated impacts of this warming trend on various aspects relevant to climate displacement.

Rising Seas: Swallowing Coastlines and Displacing Millions

With a projected rise in global sea levels by up to 0.4 meters (or more) by 2100, low-lying coastal areas, river deltas, and small island nations face an existential threat (Intergovernmental Panel on Climate Change, 2021). Entire communities could be submerged, displacing tens of millions of people. Densely populated countries like Bangladesh, Egypt, and Vietnam are particularly vulnerable, as are small island nations in the Pacific and the Caribbean (The World Bank, 2019).

Imagine the plight of residents in Bangladesh, where millions live in the Ganges Delta, a region highly susceptible to rising sea levels. Saltwater intrusion could render agricultural land unusable, displacing farmers and jeopardizing food security. Furthermore, storm surges and coastal erosion could inundate villages and displace entire communities, forcing them to migrate inland in search of higher ground (Rahman & Salehin, 2018).

Intensifying Weather Extremes: A Year-Round Displacement Threat

A 2°C temperature rise does not just mean rising sea levels; it also signifies a world with more frequent and severe weather events. Cyclones, hurricanes, and typhoons are projected to become more intense, while extreme rainfall events will lead to devastating floods (Knutson et al., 2019). Regions like South Asia, Southeast Asia, parts of the United States, Central America, and the Caribbean are particularly at risk.

These intensified weather extremes will create a year-round displacement threat. Millions could be displaced annually due to disasters like floods and cyclones, destroying homes, livelihoods, and infrastructure. The unpredictable nature of these events would make long-term planning challenging, further exacerbating the displacement crisis.

Agricultural Disruption and Desertification: When the Land No Longer Sustains

Climate change is not just about rising sea levels and extreme weather; it also disrupts agricultural production. A 2°C temperature rise is expected to significantly reduce yields for major crops like wheat and maize (Intergovernmental Panel on Climate Change, 2021). This will exacerbate food insecurity, destabilize economic growth in agricultural regions, and force people to abandon their land for sustenance.

Sub-Saharan Africa, parts of South Asia, and the Middle East are particularly vulnerable to agricultural disruption. These regions are already struggling with water scarcity, and a changing climate will only worsen their plight. Over the past few decades, millions have been displaced due to non-viable agricultural conditions and water scarcity, and this trend is projected to accelerate significantly with a two-degree Celsius temperature rise.

The Human Cost of Inaction

The estimated impacts of a two-degree Celsius temperature rise paint a grim picture. Millions will be displaced due to rising sea levels, weather extremes, and agricultural disruptions. These numbers represent real people, families, and communities uprooted from their homes and forced to rebuild their lives from scratch.

The urgency of addressing climate change must be addressed. By transitioning to renewable energy sources, adopting sustainable practices, and investing in climate adaptation strategies, we can mitigate the worst impacts of climate change and minimize displacement. The future of millions depends on our collective action today.

Ecosystem Disruption and Lost Livelihoods

Climate change is not just about physical changes to our planet; it also disrupts the delicate balance of ecosystems. A 2°C temperature rise will trigger biodiversity loss, forest die-offs, and severe impacts on freshwater and marine ecosystems (Intergovernmental Panel on Climate Change, 2021). These ecological disruptions will have cascading effects, indirectly displacing communities that depend on natural resources for their livelihoods.

The Amazon Basin, home to 10% of the world's known biodiversity, is particularly vulnerable. Climate change threatens to disrupt rainfall patterns and increase the risk of wildfires, jeopardizing the livelihoods of millions who rely on the Amazon's resources for fishing, hunting, and forestry (Amazon Conservation Association, 2023).

Similarly, coral reefs worldwide, including the iconic Great Barrier Reef, face bleaching and death due to rising water temperatures. This has severe consequences for coastal communities dependent on fishing and tourism for their income, potentially forcing them to migrate in search of new opportunities. Boreal forests, vital for carbon sequestration and home to indigenous communities, are also at risk from climate-induced changes, potentially displacing traditional inhabitants.

Economic Downturn: A Multiplier Effect on Displacement

The economic impacts of climate change will exacerbate the displacement crisis. Areas severely affected by rising sea levels, extreme weather events, and ecosystem disruptions will experience economic contraction (The World Bank, 2019). This decline in economic activity will disproportionately impact developing countries already facing poverty and limited resources.

As local economies suffer, people will be forced to migrate for better opportunities. These "economic migrants" are distinct from those directly displaced by environmental disasters but are nonetheless driven from their homes by the secondary effects of climate change. The economic fallout from climate change can potentially create a significant new wave of displacement, particularly from developing countries with limited social safety nets.

What Are the Strategies for a More Resilient Future?

The scale of the displacement challenge necessitates a multi-pronged approach. Here are some key strategic considerations:

Adaptation Strategies: Investing in infrastructure that can withstand extreme weather events, developing drought-resistant crops, and improving water management practices are crucial for enabling communities to adapt to a changing climate and minimize displacement (United Nations Framework Convention on Climate Change, 2023).

Migration Policies: Developing planned relocation strategies for communities in high-risk zones, strengthening the rights and support systems for internally displaced persons, and revising international frameworks to recognize climate refugees are all essential steps toward managing displacement effectively (Internal Displacement Monitoring Centre, 2023).

Mitigation Efforts: The most effective way to prevent large-scale displacement is to mitigate climate change. The Paris Agreement targets a global commitment to reduce greenhouse gas emissions and keep global warming below two degrees Celsius, ideally to 1.5 degrees Celsius.

Meeting these targets requires a rapid transition to renewable energy sources and an international commitment to sustainable practices (United Nations Framework Convention on Climate Change, 2015).

By implementing these strategies, we can build a more resilient future where displacement is minimized, ecosystems are protected, and communities have the necessary resources to adapt to a changing climate.

A Call to Action:

The climate crisis displaces millions, and the future holds the potential for even more significant upheaval. However, this does not have to be the inevitable outcome. By taking decisive action, we can mitigate the impacts of climate change, minimize displacement, and build a more resilient future for all. This section will explore three critical areas for action: mitigating climate change, advocating for improved policies, and individual and community-based solutions.

Mitigating Climate Change: The Root Cause

The most effective way to address climate displacement is to tackle its root cause - climate change itself. This requires a global shift toward a low-carbon future. Here are some key strategies:

Transitioning to Renewable Energy: Phasing out fossil fuels and investing in renewable energy sources like solar, wind, and geothermal power is crucial for reducing greenhouse gas emissions. This transition requires significant investment in clean energy infrastructure and government incentives to encourage widespread adoption.

Enhancing Energy Efficiency: Improving energy efficiency in buildings, industries, and transportation sectors can significantly reduce energy consumption and associated emissions. This includes promoting energy-efficient appliances, buildings, and industrial processes.

Protecting Forests: Forests play a vital role in carbon sequestration. Implementing effective forest conservation and restoration programs can significantly reduce atmospheric CO2 levels.

These mitigation efforts will reduce future displacement and create economic opportunities in green technologies and sustainable practices.

Advocating for Improved Policies:

While mitigating climate change is essential, we must address the displacement and prepare for what will come. This requires improved policies at the national and international levels.

Developing a Legal Framework for Climate Refugees: The international community must work toward a comprehensive legal framework to protect climate refugees. This framework should address issues like access to safe havens, resettlement opportunities, and humanitarian assistance.

Supporting Adaptation Efforts in Vulnerable Regions: Investing in climate-resilient infrastructure, early warning systems, and livelihood diversification programs in vulnerable regions can help communities adapt to changing environmental conditions and minimize displacement risks.

Strengthening Local Governance: Empowering local communities to manage their resources and develop adaptation strategies is crucial. This includes providing them with the financial resources, technical expertise, and decision-making power to respond effectively to climate challenges.

Advocating for these policy changes can help individuals and organizations play a vital role in shaping a global response to climate displacement.

Individual and Community Action:

Climate action is not limited to large-scale initiatives. Individuals and communities can contribute significantly through everyday choices and local initiatives.

Reducing Your Carbon Footprint: Making conscious choices to reduce your carbon footprint, such as using public transportation, conserving energy at home, and consuming less meat, can collectively reduce greenhouse gas emissions.

Supporting Sustainable Practices: Choose sustainable products, support businesses committed to environmental responsibility, and advocate for climate-friendly policies in your community. Every action, however small, sends a powerful message.

Engaging in Community Initiatives: Get involved in local organizations working on climate solutions, participate in environmental education programs, and raise awareness about climate displacement in your community.

Working together, individuals and communities can build collective resilience, promote sustainable practices, and hold decision-makers accountable for climate action.

The climate crisis is a complex challenge, but it is not insurmountable. By adopting a multi-pronged approach that combines ambitious mitigation strategies, improved legal frameworks, and individual and community action, we can build a more resilient future where displacement is minimized, and communities can adapt to a changing climate. The time for action is now. Let us rise to the challenge and ensure a safer planet for future generations.

Collective Action for Climate Refugees

Climate change displaces millions worldwide, creating a profound humanitarian crisis. Climate refugees face a multitude of challenges, including losing their homes and livelihoods, struggling to access necessities like food and water, and grappling with the emotional and psychological trauma of displacement. The international legal framework is inadequate to address their needs, and the future holds the potential for even greater displacement if we fail to act decisively.

However, amid this bleak reality, there exists a glimmer of hope. By acknowledging the scale of the challenge and taking proactive steps at the global and local levels, we can build a more resilient future for climate refugees and all of humanity.

CHAPTER 7
Innovating Our Way Forward

Technology and innovation have emerged as potent tools in addressing climate challenges. They illuminate groundbreaking solutions and their potential for fostering a sustainable future. These advancements pave the way for transformative changes, offering hope amidst environmental concerns.

From renewable energy sources to carbon capture technologies, the innovation landscape underscores a commitment to combating climate change. Integrating these pioneering solutions becomes increasingly imperative as society navigates the complexities of a warming planet.

Let us learn how to set the stage for exploring the intersection of technology, innovation, and environmental stewardship in pursuing a greener tomorrow.

Renewable Energy

Renewable energy is derived from natural sources and replenished faster than consumed (NRDC, 2023). Unlike fossil fuels, which take millions of years to form, renewable resources are constantly being renewed, making them a sustainable and environmentally friendly alternative.

Renewable energy is crucial in mitigating climate change and enhancing energy resilience (by us, 2023).

Fossil fuels, the primary source of traditional energy, release greenhouse gasses like carbon dioxide when burned, contributing to global warming. In contrast, renewable energy sources like solar, wind, and geothermal produce minimal to no emissions, helping to reduce our carbon footprint and slow climate change (UN, 2023).

Renewable energy also strengthens energy resilience by diversifying our energy portfolio. Our dependence on finite fossil fuels leaves us vulnerable to price fluctuations and geopolitical instability. Investing in renewable sources can create a more secure and reliable energy system less susceptible to external disruptions (Byjus, 2023).

Harnessing the Power of Nature:

The sun's brilliance offers a vast and ever-present source of clean energy. Photovoltaic (PV) panels act as the sun's collectors, transforming its rays into electricity that can power homes, businesses, and entire communities [NRDC, 2023]. Solar energy's versatility makes it a significant player in transitioning toward a more sustainable future.

Wind, another natural phenomenon, can be harnessed to generate electricity through wind turbines. These elegant structures capture the kinetic energy of moving air, converting it into a usable power source.

Wind farms and clusters of wind turbines are becoming an increasingly common sight, particularly in regions blessed with consistent wind patterns (Byjus, 2023).

Deep within the Earth lies another source of renewable energy: geothermal heat. This thermal energy, emanating from the planet's core, can be utilized for two primary purposes: generating electricity and heating buildings (TWI-Global, 2023). Geothermal energy offers a unique advantage – its reliability. Unlike solar and wind power, which can fluctuate with weather conditions, geothermal plants provide a consistent baseload of power, making them a valuable asset to any energy grid.

Harnessing the power of the sun, the wind, and the Earth's internal heat can help us achieve a clean and sustainable energy future. These renewable resources offer a promising path toward a more secure and environmentally responsible energy landscape.

How Innovation Can Save the Planet?

Climate change is the defining challenge of our generation, and it can feel overwhelming. However, here is the good news: I am a huge believer in the power of human ingenuity to tackle even the most challenging problems. That is where technology and innovation come in – they are not just cool gadgets but the tools we need to build a sustainable future.

Think about it – the very problems technology helped create, like fossil fuel reliance, can be solved with even more innovative solutions. For example, the rise of AI is revolutionizing how we approach climate action. AI-powered models are used to develop early warning systems for natural disasters, a game-changer for protecting communities ([UNFCC], 2023). It also helps optimize energy grids by predicting demand, leading to reduced emissions from a more efficient system (IEA, 2023).

The recent push for innovation is truly inspiring. International agreements like the Paris Agreement set ambitious climate targets, and the race is on to meet them. We are seeing a surge in critical technologies like advanced battery storage, which is crucial for the large-scale adoption of renewable energy sources like solar and wind (IRENA, 2023). Plus, imagine a world where carbon capture and storage (CCS) technologies become more efficient and affordable, allowing us to remove existing greenhouse gasses from the atmosphere. (Global CCS Institute, 2023)

These are just a few examples, and the field of climate tech is constantly evolving. It is exciting to think about the possibilities – new materials, cleaner industrial processes, and maybe even technologies we have not even dreamed of yet.

There is no silver bullet, but with continued investment, collaboration, and a commitment to innovation, technology can be the key to unlocking a sustainable future for our planet.

The Power of Personal Action: A Case for Renewable Energy at Home

Beyond the global picture, individual actions also play a crucial role in promoting a sustainable future. Let us explore a personal experience that exemplifies the power of renewable energy at the household level.

Investing in Solar Energy

I decided to install a solar energy system for my home, investing over $50,000. This choice was driven by a twofold desire: to minimize my environmental impact and significantly reduce my monthly electricity bills. Before the installation, my monthly household electricity costs averaged around $300. However, the switch to solar power yielded dramatic results. My electricity bill plummeted, dropping as low as an

$8.00 credit in one month. Any excess energy the system generates feeds back into the grid, even earning a modest compensation.

Sizable federal tax incentives, covering approximately one-third of the total cost, significantly eased the financial burden of the initial investment. This government support program highlights the crucial role policymakers can play in encouraging the adoption of renewable energy technologies.

Smart Water Heaters

The second major energy consumer in my home was the water heater. My old unit was an inefficient model that ran constantly, consuming unnecessary energy. While I initially considered purchasing a new, "intelligent" water heater, their cost proved a deterrent.

Fortunately, I discovered a more cost-effective solution – a retrofit device that transformed my conventional "dumb" water heater into a smart one. This ingenious device lets me control the water heating schedule through my smartphone. Now, I can strategically heat water only during peak sunlight hours, typically from 11:00 AM to noon, maximizing the utilization of my solar energy generation.

Despite this limited heating window, the hot water remains readily available for showers, laundry, and dishwashing. This simple and affordable upgrade, costing around $150, resulted in significant energy savings. With such a short payback period measured in weeks, this demonstrates the potential for intelligent technologies to empower individuals to make sustainable choices without incurring a high upfront cost.

This personal experience serves as a microcosm of renewable energy's broader potential. By embracing these technologies, individuals can reduce their environmental footprint and energy bills and contribute to a cleaner and more sustainable future for future generations.

Sustainable Agriculture Techniques

For too long, conventional agriculture has damaged our environment. Nevertheless, a growing movement is toward a more responsible approach—sustainable agriculture. This is not just about trendy buzzwords; it is about nurturing the land that sustains us and creating a healthy food system for people and the planet.

Sustainable agriculture offers many benefits. It starts with the very foundation of our food system—healthy soil. By employing crop rotation and reducing tillage, sustainable agriculture promotes increased organic matter content, improves nutrient cycling, and prevents soil erosion (FAO, 2023). This translates to healthier crops and, ultimately, more nutritious food.

However, the benefits extend far beyond the soil. Sustainable agriculture fosters biodiversity by creating beneficial insect and pollinator habitats (World Wildlife Fund, 2023). This diversity strengthens ecosystems and makes them more resilient to pests and diseases. This reduces the need for harmful chemical pesticides, contributing to a safer food supply and a healthier environment.

The most crucial benefit of sustainable agriculture is its role in ensuring food security for future generations. By protecting soil health and promoting biodiversity, these practices ensure long-term agricultural productivity. Additionally, by minimizing water usage and reducing greenhouse gas emissions, sustainable agriculture helps mitigate climate change, a growing threat to global food security (The World Bank, 2023).

A Toolbox for Sustainability

There is no one-size-fits-all approach to sustainable agriculture, but several vital practices can make a significant difference. Crop rotation, for instance, involves planting different crops in the same field over time. This helps maintain soil fertility, breaks pest and disease cycles, and promotes a more diverse and resilient ecosystem (Cornell University, College of Agriculture and Life Sciences, 2023).

Another powerful tool is agroforestry, which integrates trees and shrubs into agricultural landscapes. Trees provide shade and shelter for crops, improve soil quality, and offer additional income sources through fruits, nuts, or timber (World Agroforestry, 2023).

Finally, Integrated Pest Management (IPM) offers a strategic approach to managing pests. Instead of relying heavily on chemical pesticides, IPM focuses on natural control methods, like attracting beneficial insects or using biological controls. This reduces agriculture's environmental impact and promotes long-term pest control solutions (Michigan State University, Integrated Pest Management Program, 2023).

By embracing these practices and continuously innovating, we can cultivate a future where agriculture is productive and sustainable. It is a future where healthy soil nourishes healthy food, biodiversity thrives, and food security is a reality.

Actionable Insights for Farmers and Policymakers

We have explored the benefits of sustainable agriculture and the essential practices that underpin it. However, how can we translate this knowledge into action? Here are some actionable insights for both farmers and policymakers to embrace a more sustainable future:

Farmers: Embracing Agroecology

Agroecology is a holistic approach to farming that mimics natural ecosystems. It prioritizes crop rotation, cover cropping, and integrated pest management (IPM). By adopting these techniques, farmers can:

Enhance soil health: Agroecological practices promote organic matter content, improve nutrient cycling, and prevent erosion, leading to healthier and more productive crops (FAO, 2023).

Increase biodiversity: Diverse landscapes attract beneficial insects and pollinators, making an ecosystem more resilient and less susceptible to pests and diseases (World Wildlife Fund, 2023).

Reduce reliance on external inputs: Agroecological practices often minimize the need for chemical fertilizers and pesticides, leading to lower costs and a reduced environmental footprint.

Policymakers: Fostering Innovation and Collaboration

Policymakers play a crucial role in supporting the transition to sustainable agriculture. Here are some key actions they can take:

Investing in research and education: Funding research on agroecological practices and providing training programs for farmers are essential to equip them with the knowledge and skills necessary for success (The World Bank, 2023).

Implementing policy incentives: Incentive programs like subsidies for organic farming and financial assistance for implementing sustainable practices can encourage widespread adoption (European Commission, 2023).

Promoting collaboration: Building bridges between farmers, researchers, and policymakers fosters knowledge exchange and best practices, accelerating progress toward a more sustainable food system.

The transition to sustainable agriculture requires a collaborative effort. Farmers can share their experiences and challenges, while policymakers can work with researchers to develop effective support systems. Ultimately, working together can create a future where agriculture is productive, environmentally responsible, and socially just.

Sustainable agriculture is not just about changing how we farm; it is about changing how we think about food. It is about recognizing the interconnectedness of our planet and our responsibility to future generations. By embracing these actionable insights, farmers and policymakers can play a critical role in building a more sustainable and resilient food system.

Waste No More: Waste-to-Wealth Solutions

Our current throw-away culture generates a staggering amount of waste, posing a significant environmental threat. However, what if we could see waste not as a burden but as a hidden treasure? The concept of waste-to-wealth offers a revolutionary approach, transforming discarded materials into valuable resources.

Waste-to-wealth transcends mere technological innovation; it represents a paradigm shift toward a win-win-win scenario for our planet, prosperity, and resourcefulness. By diverting waste from landfills and incineration, we demonstrably reduce greenhouse gas emissions and environmental contamination. Envision a future where responsible waste management is the norm, with cleaner air, healthier soil, and pollution-free waterways.

Furthermore, waste-to-wealth technologies unlock a wellspring of economic opportunities. From engineers pioneering innovative recycling facilities to entrepreneurs establishing robust waste collection networks, these solutions fuel economic growth and empower local communities.

Finally, waste-to-wealth embodies the core principles of a circular economy. This model prioritizes keeping resources in use for extended periods, shunning the traditional linear "take-make-dispose" approach. Waste becomes a valuable input, continuously recycled and repurposed to minimize waste generation and maximize resource efficiency. In essence, waste-to-wealth fosters a more sustainable future for generations to come.

Transforming Waste with Cutting-Edge Technologies

The world of waste management is undergoing a significant transformation. No longer are we simply discarding unwanted materials; a new era of "waste-to-wealth" technologies is emerging, offering innovative solutions that address our growing waste problem and generate valuable resources.

Advanced Gasification stands at the forefront of this movement. This process takes organic waste, such as food scraps and yard trimmings, and converts it into syngas, a highly versatile fuel source. Syngas can be utilized in many ways: generating electricity, creating biofuels, or even producing the chemicals that make up everyday products like plastics and detergents. A prime example is the Enerkem facility in Edmonton, Canada, which serves as a beacon of progress. This facility demonstrates the immense potential of advanced gasification by diverting thousands of tons of waste from landfills and transforming them into clean, renewable energy.

Anaerobic Digestion offers another powerful approach. This technology harnesses the power of nature's decomposers, microorganisms, to break down organic waste in an oxygen-free environment. The resulting product is biogas, a clean-burning fuel that can power vehicles and generate electricity. Furthermore, the process yields digestate, a nutrient-rich fertilizer that can improve soil health. The Aarhus University project in Denmark exemplifies the potential of anaerobic digestion, showcasing how food waste can be transformed into valuable resources that contribute to a sustainable future.

Plastic pollution presents a particularly pressing challenge. However, innovative solutions are emerging to tackle this issue. Pyrolysis stands out as a promising approach. This process utilizes heat to break down plastic waste without oxygen, converting it into liquid fuels such as diesel or gasoline. Plastic Energy's plant in Seville, Spain, demonstrates the viability of pyrolysis, offering a second life for plastic waste and paving the way for a more sustainable future.

Chemical Recycling takes waste management to a whole new level. This cutting-edge technology breaks down plastic waste into its fundamental building blocks, monomers. These monomers can then be used to create new, virgin-quality plastics. BASF's Chem Cycling project in Germany exemplifies the potential of chemical recycling. This project paves the way for a more circular economy by offering a closed-loop solution for plastic waste management.

These advancements represent just a glimpse into the exciting world of waste-to-wealth technologies. As this field continues to evolve, we can expect even more efficient and sustainable solutions to emerge, ushering in a future where waste becomes a resource rather than a burden.

Waste-to-wealth is not just a technological marvel but a call to action. It is about fostering a society that values resource efficiency and prioritizes responsible waste management. By embracing these innovative solutions and adopting more sustainable practices, we can create a future where waste becomes a resource, not a burden.

Electric Vehicle (EV) Market and Its Impact on Climate

As someone who profoundly values sustainability and the shift toward renewable energy, the rapid growth of the electric vehicle (EV) market over the past decade has particularly excited me. Remarkable technological advancements, progressive government policies, and increasing global awareness about climate change have driven this surge. Let us examine the current state of the EV market, its anticipated growth, and how this transition impacts our climate. No company epitomizes this more than Elon Musk's TESLA.

Current State of the EV Market: The global EV market has taken off. Just look at the numbers: sales skyrocketed from 2.1 million units in 2019 to 10.2 million in 2022. It is incredible to see China, Europe, and the United States leading the charge, collectively making up over 80% of global EV sales. Big names like Tesla, Nissan, General Motors, and Volkswagen heavily invest in EV technology and infrastructure, propelling the market forward.

The Chinese EV market, however, may face new challenges in the European Union (EU). The EU announced potential tariffs on Chinese electric vehicles, labeling them a threat to its industry. Chinese EV manufacturers could face tariffs if discussions with Chinese authorities do not yield a solution by July 4. An ongoing EU investigation claims that an influx of cheap, government-subsidized Chinese cars harms the trade bloc. China's response has been critical, alleging that the tariffs violate international trade rules and calling the investigation "protectionism." (Labiak, 2024)

Governments worldwide are pivoting in this transition. They are rolling out various incentives to boost EV adoption, from tax credits and subsidies to rebates. Policies like the European Union's Green Deal and China's New Energy Vehicle (NEV) program are designed to cut greenhouse gas emissions and speed up the move to electric mobility. Countries like the UK, France, and Norway plan to phase out internal combustion engine (ICE) vehicles by 2030-2040.

The EU's potential tariffs reflect this policy approach. Chinese EV makers who cooperated with the investigation initiated by the European Commission in October may face an average 21% duty, while those who did not cooperate might see a 38.1% tariff. Specific charges will apply to companies like BYD, Geely, and SAIC, with non-Chinese companies producing EVs in China, including EU-based ones like BMW, also affected. (Labiak, 2024)

Technology is another driving force. Advances in battery technology have significantly increased the range and performance of EVs while reducing costs. Fast-charging infrastructure has made EVs much more convenient, addressing one of the main hurdles to their adoption. Plus, innovations in autonomous driving and intelligent grid integration are set to enhance EVs' appeal and efficiency further.

Projected Growth of the EV Market: The future looks bright for the EV market. The International Energy Agency (IEA) projects that there will be 145 million EVs on the road by 2030, making up 7% of the global vehicle fleet. EV sales are expected to grow at an impressive compound annual growth rate (CAGR) of 29%, driven by falling battery costs, robust government support, and rising consumer demand.

To support this growth, significant investments in charging infrastructure are crucial. Both public and private sectors are heavily investing in expanding charging networks, with companies like ChargePoint, Electrify America, and Tesla leading the way globally.

One of the most exciting aspects of EVs is their potential to reduce greenhouse gas emissions drastically. EVs produce zero tailpipe emissions, offering a significant advantage over ICE vehicles. The IEA estimates widespread EV adoption could cut global CO2 emissions by 1.5 gigatons annually by 2030. As we shift toward renewable energy sources for electricity, the environmental benefits of EVs will only grow.

EVs also improve air quality, especially in densely populated urban areas. Reducing tailpipe emissions leads to substantial decreases in pollutants like nitrogen oxides (NOx) and particulate matter (PM), which harm our health.

It is essential to consider the entire lifecycle of EVs. While they produce fewer emissions during operation, producing batteries and sourcing raw materials have environmental impacts. Advances in battery recycling and sustainable sourcing are essential to minimize the overall ecological footprint of EVs. Companies and governments are investing in research and development to improve battery recycling technologies and create a circular economy for EV batteries.

In summary, The EV market is on the brink of substantial growth, fueled by technological advancements, supportive government policies, and a growing global awareness of climate change. Widespread EV adoption has the potential to significantly reduce greenhouse gas emissions, improve air quality, and support international efforts to combat climate change. However, it is crucial to address the environmental impacts of battery production and recycling to maximize the climate benefits of EVs. Continued investment in infrastructure, technology, and sustainable practices will be essential for the EV market to reach its full potential as a critical driver of a cleaner, greener future.

The EU's proposed tariffs on Chinese EVs underscore the complex dynamics of international trade and the balance between protecting domestic industries and promoting free trade. As global leaders navigate these challenges, the focus remains on fostering a robust and sustainable EV market that benefits the environment and the global economy. (Labiak, 2024)

Successful Projects and Emerging Trends

Waste-to-wealth is no longer confined to scientific papers and pilot projects. Cities around the world are implementing innovative solutions and achieving impressive results. Here are a few examples that showcase the power of this approach in action:

Zero-Waste Policies: San Francisco, USA, has set a bold goal – achieving zero waste by 2030. Through composting programs, recycling initiatives, and public education campaigns, the city has reached an impressive diversion rate of 80%, meaning that 80% of its waste is diverted from landfills (San Francisco Environment, 2023). This is a testament to the effectiveness of comprehensive waste management strategies.

Innovative Waste Management: Technology is crucial in optimizing waste collection and processing. Seoul, South Korea, has embraced this concept by implementing a network of "smart bins." These bins use sensors to monitor fill levels and communicate with waste collection trucks, optimizing routes and reducing unnecessary fuel consumption (World Economic Forum, 2023). This is just one example of how the Internet of Things (IoT) revolutionizes waste management.

Blockchain for Transparent Waste Management: Blockchain technology offers a secure and transparent platform for tracking waste streams and ensuring responsible waste management practices (IBM, 2020). By creating an immutable record of waste collection, processing, and recycling, blockchain can improve traceability, reduce fraud, and incentivize sustainable waste management practices throughout the supply chain. Pilot projects are already underway in cities like Rotterdam, Netherlands, where blockchain is being used to track plastic waste from collection points to recycling facilities, promoting accountability and transparency in the waste management process (IBM, 2020). This technology can potentially revolutionize the waste-to-wealth sector by fostering stakeholder trust and collaboration, ultimately contributing to a more circular economy.

Urban Upcycling Initiatives: Reusing and repurposing waste materials is another exciting trend gaining traction. Barcelona, Spain, is a frontrunner in this space with its "The Fab City" initiative. This program connects architects, designers, and entrepreneurs to create innovative products and infrastructure from discarded materials. From transforming construction waste into furniture to using old tires for playgrounds, "The Fab City" demonstrates the potential of urban upcycling to create a more circular economy (The Fab City, 2023).

These successful projects offer valuable lessons for cities around the world. Investing in innovative technology, implementing intelligent policies, and fostering a culture of upcycling can contribute to the waste-to-wealth revolution.

The transition to a more sustainable future requires a collaborative effort. Governments, businesses, and individuals all have a role to play. Governments can create supportive policies and invest in infrastructure, companies can develop innovative technologies and circular economy models, and individuals can make conscious choices to reduce waste and support sustainable products and services.

By working together, we can transform waste from a burden into a valuable resource, creating a cleaner, healthier, and more prosperous future for all.

A Call to Action

We have explored the transformative power of technology and innovation in addressing climate change and achieving the Sustainable Development Goals (SDGs). Innovation provides us the tools to build a better future, from renewable energy sources and sustainable agriculture techniques to waste-to-wealth solutions.

Imagine a world powered by clean energy, where solar panels grace rooftops and wind turbines dot landscapes, generating electricity without harmful emissions. Imagine fields brimming with healthy crops nurtured by regenerative agriculture practices that promote soil health and biodiversity. Imagine a world where waste becomes a resource, transformed into valuable products through innovative technologies. This is the future that innovation promises, where we live in harmony with our planet.

The SDGs provide a comprehensive roadmap for achieving a sustainable and prosperous future for all. These goals address critical issues like poverty, hunger, clean water, and climate change. Innovation is essential in achieving these goals by developing new technologies, creating sustainable practices, and fostering collaboration across sectors.

The journey toward a sustainable future requires a collective effort. Governments can set ambitious goals, businesses can invest in sustainable solutions, and individuals can make conscious choices in their everyday lives. However, the key to achieving these goals lies in relentless innovation. We need to continuously push the boundaries of technology, explore new approaches, and adapt to evolving challenges.

The time to act is now. This book has only scratched the surface of the transformative power of technology and innovation in addressing climate change and achieving the SDGs. We have the knowledge, the tools, and the collective will to create a more sustainable future.

Do not wait for someone else to take the lead. Explore the possibilities – learn about renewable energy sources, sustainable agriculture practices, and waste-to-wealth solutions. Make informed choices in your everyday life. Support businesses committed to sustainability. Let us create the future we want –powered by clean energy, nurtured by regenerative agriculture, and fueled by innovation.

This call to action empowers individuals to take the first step. We can pave the way for a collective movement toward a brighter future by igniting a passion for sustainability and encouraging further exploration.

CHAPTER 8
The Politics Of Climate: Navigating Policy And Power

"An African proverb says, 'If you want to go quickly, go alone. If you want to go far, go together.' We need to go far, quickly. We must abandon the conceit that individual, isolated, private actions are the answer. They can and do help. However, they will not take us far enough without collective action."

— Al Gore

The opening quote by Al Gore succinctly captures the essence of this chapter: We cannot tackle climate change alone. Individual actions, while commendable, are not enough. We require a collective effort to achieve lasting change—a united front where governments, businesses, and individuals work together toward a sustainable future.

The previous chapter explored the exciting world of renewable energy and technological advancements that empower us to combat climate change. This chapter will plunge into the crucial role of stakeholders – the various entities with a vested interest in the environment – and how their collaborative action forms the bedrock of a successful climate strategy.

Governments wield immense power in shaping how we address climate change. They can enact laws and regulations that significantly reduce carbon emissions and mitigate environmental damage. For instance, carbon taxes implemented by governments incentivize businesses and individuals to shrink their carbon footprints (International Monetary Fund, 2020). International agreements, like the Paris Agreement, are crucial for coordinated global action, setting ambitious emissions reduction targets, and fostering collaboration among nations (United Nations Framework Convention on Climate Change, 2015).

A. Governments

Governments hold immense power in shaping climate change policies and actions. Their ability to enact laws and regulations positions them as key drivers in reducing carbon emissions and mitigating environmental impacts. Here is a closer look at how governments are wielding their influence:

Regulatory Frameworks:

Governments are uniquely positioned to influence climate policy through legislation and regulation. They establish regulatory frameworks such as carbon pricing, emissions trading systems, and renewable energy mandates, which are critical tools in this fight (Environmental Protection Agency, 2021). Carbon pricing mechanisms, including carbon taxes and cap-and-trade systems, incentivize businesses to reduce greenhouse gas emissions by assigning a cost to their carbon footprint. The European Union Emissions Trading System (EU ETS) is a prime example. It sets a cap on the total amount of greenhouse gases that can be emitted and allows trading emission allowances (European Commission, n.d.).

Subsidies and Incentives:

Governments do not just regulate; they also incentivize. They offer subsidies and incentives to promote renewable energy and energy efficiency. Subsidies for wind, solar, and other renewable energy sources can make these technologies more competitive with fossil fuels (International Renewable Energy Agency, 2023). Tax credits for energy-efficient buildings and electric vehicles also nudge individuals and businesses to adopt greener technologies. The United States, for example, provides federal tax credits for solar panel installations through the Investment Tax Credit (ITC), which has significantly boosted the solar industry (Solar Energy Industries Association, 2024).

International Agreements:

The Paris Agreement is a beacon of international cooperation on the global stage. Adopted in 2015, it brings all nations together to undertake ambitious efforts to combat climate change and adapt to its effects (United Nations Framework Convention on Climate Change, 2015). Each country sets its own nationally determined contributions (NDCs) to mitigate climate change, aiming to limit global warming to well below 2 degrees Celsius above pre-industrial levels and to pursue efforts to limit the temperature increase even further to 1.5 degrees Celsius.

Public Pressure and Advocacy:

Public opinion and advocacy are not bystanders; they play a significant role in shaping government policies. Grassroots movements, protests, and lobbying efforts can influence political leaders to prioritize climate action. As mentioned in Chapter 4, movements like Fridays for Future, initiated by Greta Thunberg, have mobilized millions worldwide, particularly the youth, to demand more decisive climate action from their governments (Fridays for Future, n.d.). This public pressure can lead to more ambitious climate policies and commitments.

B. Businesses

Businesses, giant corporations, have a significant environmental impact. To become responsible stewards, they can adopt sustainable practices like utilizing renewable energy sources, minimizing waste, and improving energy efficiency. Transparency is critical; companies must implement these eco-friendly practices and openly report their progress (World Business Council for Sustainable Development, 2023). By doing so, they can lead by example and influence their industries to embrace a more sustainable future.

Corporations Take Responsibility:

Businesses have a significant environmental footprint, and corporate responsibility requires actively reducing it. This means adopting sustainable practices encompassing switching to renewable energy sources, minimizing waste generation, improving energy efficiency throughout operations, and slashing greenhouse gas emissions. Tech giants like Apple and Google set a benchmark for their industry by pledging to run entirely on renewable energy (Apple, 2022; Google, 2022).

Sustainable Supply Chains:

Sustainable supply chain management is another critical area. Businesses can shrink their carbon footprint by sourcing materials sustainably, minimizing transportation emissions, and ensuring their suppliers adhere to environmental standards. For instance, Walmart's Project Gigaton initiative actively works to eliminate one billion metric tons of greenhouse gases from its global supply chain by 2030 (Walmart, 2023).

Transparency and Reporting:

Transparency in reporting environmental impact is vital for holding businesses accountable. Standards and frameworks like the Global Reporting Initiative (GRI) and the Task Force on Climate-related Financial Disclosures (TCFD) provide clear guidelines for companies to disclose their environmental performance and climate-related risks (Global Reporting Initiative, n.d.*; Task Force on Climate-Related Financial Disclosures, n.d.*). Investors increasingly use this information to make informed investment decisions, driving a stronger emphasis on sustainability within business operations.

Innovation for a Greener Future

Innovation is a powerful engine for corporate climate action. Businesses investing in green technologies research and development can unlock breakthroughs that significantly reduce environmental impact. For example, Tesla's advancements in electric vehicle technology have accelerated the shift away from gasoline-powered cars, promoting cleaner transportation options worldwide (Tesla, 2024).

C. Individuals:

Regular people can significantly influence climate policies through advocacy and voting. Individuals can drive substantial changes by actively supporting leaders prioritizing environmental issues and participating in climate advocacy efforts. Voting for climate-conscious candidates ensures that climate change policies are prioritized and implemented. Public pressure, fueled by grassroots movements, is a powerful tool to push governments and businesses to take meaningful action.

Advocacy and Activism in Action

Individuals can influence climate policy through advocacy and activism. They can join demonstrations, sign petitions, and connect with environmental organizations like Greenpeace or the Sierra Club (Greenpeace International, n.d.; *Sierra Club, n.d.*). These organizations provide platforms for individuals to engage in environmental advocacy, pushing for local, national, and international policy changes.

Political Engagement for Climate Solutions

Voting for candidates who prioritize climate action is an influential way individuals can influence government policies. Climate change is a central issue in many political campaigns, and electing leaders committed to addressing climate issues can lead to more robust environmental policies. Do not stop voting! Attend town hall meetings and communicate your climate concerns to elected officials to enhance your impact.

Sustainable Choices for a Greener Future

Making sustainable lifestyle choices can also contribute to climate mitigation. Reduce energy consumption, use public transportation whenever possible, consider adopting a plant-based diet, and minimize waste generation. These practical steps, taken by many, can significantly reduce greenhouse gas emissions. The zero-waste movement, for example, encourages individuals to minimize their waste production, promoting a more sustainable way of living (Zero Waste International Alliance, n.d.).

D. International Cooperation

Global challenges like climate change demand international cooperation. Agreements like the Paris Agreement are crucial in coordinating efforts across countries. These agreements establish emission reduction targets and facilitate the sharing of technology and resources, enabling countries to combat climate change effectively. International cooperation ensures national policies align with global climate goals, fostering a unified and robust response to this crisis.

Global Agreements and Frameworks:

The global nature of climate change necessitates international cooperation. Agreements like the Kyoto Protocol and the Paris Agreement establish a framework for international collaboration (United Nations Framework Convention on Climate Change, 2015). These agreements promote sharing technologies, financial resources, and best practices among countries, fostering coordinated efforts to reduce global emissions.

Financial Mechanisms:

Financial mechanisms like the Green Climate Fund (GCF) empower developing countries in their fight against climate change. The GCF actively provides funding for projects that promote low-emission and climate-resilient development (Green Climate Fund, n.d.). This financial support is critical for countries that lack the resources to invest in climate action independently.

Technology Transfer:

Technology transfer is another crucial aspect of international cooperation. Developed countries can bridge the capability gap by providing developing nations access to clean technologies and expertise (United Nations Environment Programme, 2020). This ensures that all countries can participate effectively in global climate mitigation efforts.

Monitoring and Accountability:

Global agreements also incorporate mechanisms for monitoring and accountability. Regular reporting and review processes ensure that countries are fulfilling their commitments and making progress toward their climate goals. The Paris Agreement's transparency framework requires countries to provide clear and accurate information about their emissions and measures to address climate change (United Nations Framework Convention on Climate Change, 2015).

E. Politicians Leading the Charge on Climate Change

Climate change has become a pressing issue in recent decades, and several political figures have emerged as influential leaders in addressing it. Here is a closer look at their contributions:

Al Gore, the former Vice President of the U.S.A., leveraged his platform to raise global awareness. His documentary, "An Inconvenient Truth" (2006), served as a powerful wake-up call, highlighting the urgency of climate action (Gore, 2006). Through his organization, The Climate Reality Project, Gore continues to advocate for renewable energy and policy changes to combat climate change (Climate Reality Project, n.d.).

Angela Merkel, the former German chancellor, played a pivotal role in shaping European and global climate policy. She championed ambitious climate goals and spearheaded Germany's energy transition, "Energiewende," which aims to phase out nuclear power and shift toward renewable energy sources (The Federal Ministry for Economic Affairs and Climate Action, 2023).

Xi Jinping, the current President of China, has positioned China as a significant player in global climate action. Under his leadership, China pledged to reach peak carbon emissions by 2030 and achieve carbon neutrality by 2060 (Xinhua, 2020). Additionally, China has made significant investments in developing and deploying renewable energy technologies (International Renewable Energy Agency, 2023).

Barack Obama, the former President of the United States, played a crucial role in the 2015 Paris Agreement. His administration committed the U.S. to significant emissions reductions and implemented various domestic policies to promote clean energy and reduce greenhouse gas emissions (White House, 2016). These efforts included the Clean Power Plan and new transportation standards to limit carbon pollution and encourage renewable energy investments (White House, 2016).

Narendra Modi, the Prime Minister of India, strongly emphasizes renewable energy, particularly solar power. He launched initiatives like the International Solar Alliance and committed India to significant renewable energy capacity targets. Modi's focus on solar energy has positioned India as a global leader in the fight against climate change, reflecting his commitment to sustainable development (Ministry of New and Renewable Energy, Government of India, 2023).

In contrast, recent election campaigns have not prominently featured climate change as a critical issue. The focus has shifted toward immediate concerns like geopolitical conflicts, such as the wars in Ukraine and Palestine, and domestic issues like immigration and religious sentiments. Modi's recent political strategies have centered on gaining a majority by appealing to religious and cultural sentiments, such as the inauguration of a significant temple in Varanasi, rather than emphasizing climate change (Brookings,2023).

Although Modi has been recognized for his renewable energy initiatives on the global stage, these topics remained the same during the latest elections. This suggests a disconnect between Modi's international climate commitments and domestic political agenda, recently prioritizing other pressing issues over climate action (Modi, 2023).

Emmanuel Macron, the President of France, has been a vocal advocate for climate action, especially after the U.S. withdrawal from the Paris Agreement. Macron has hosted international climate summits and pushed for more robust EU climate policies. His leadership has been crucial in maintaining global momentum toward achieving the goals set out in the Paris Agreement (The Guardian, 2017).

Greta Thunberg, a climate activist from Sweden, significantly influenced global climate discourse through her passionate activism. She inspired millions of young people worldwide and pressured political leaders to take urgent action on climate change (Schroeder, 2019). Her speeches, including the one at the UN Climate Action Summit in 2019, highlighted the lack of action by world leaders and called for immediate and substantial measures to combat climate change (Schroeder, 2019).

Ban Ki-Moon, former Secretary-General of the United Nations, played a pivotal role in promoting global climate action, particularly leading up to the Paris Agreement. His advocacy for sustainable development and climate resilience helped shape international climate policies (United Nations, n.d.).

Jacinda Ardern, Prime Minister of New Zealand, prioritized climate change in her government's policies. She committed New Zealand to carbon neutrality by 2050 and implemented measures to reduce emissions and promote renewable energy (New Zealand Government, n.d.).

Ursula von der Leyen, President of the European Commission, championed the European Green Deal, aiming to make Europe the first climate-neutral continent by 2050. She pushed for ambitious targets to reduce greenhouse gas emissions and transition to renewable energy (European Commission, n.d.).

Kofi Annan (Ghana), a former UN Secretary-General, consistently advocated urgent climate action. He emphasized the need for developed countries to support developing nations while urging emerging economies to reduce carbon emissions. Annan promoted the inclusion of civil society and the private sector in climate initiatives, as exemplified by the UN Global Compact, which guided businesses in contributing to the Millennium Development Goals (MDGs) (International & Executive Programs; UC Berkeley, n.d.). Annan also co-founded the Alliance for a Green Revolution in Africa (AGRA), focusing on climate-smart agriculture and reforestation efforts (International & Executive Programs; UC Berkeley,n.d.).

During his tenure, Annan called for global leadership on climate change, highlighting the issue in various forums and supporting policies aimed at mitigation and adaptation (The Elders, 2024). He chaired the Africa Progress Panel, which linked scientific research with advocacy to promote sustainable development (Africa Progress Group, 2018). Annan's efforts left a legacy in environmental diplomacy, influencing global climate policies through science-based advocacy and multilateral engagement (International & Executive Programs; UC Berkeley, n.d.).

These leaders significantly contributed to climate policy and action, shaping the global response to climate change through advocacy, legislation, and international agreements.

Conclusion:

Climate change is a complex challenge, but it is not insurmountable. Effectively addressing this global crisis necessitates a united effort from all stakeholders – governments, businesses, and individuals. Recognizing each group's unique power and potential contributions is fundamental to driving meaningful change toward a sustainable future.

The path to a sustainable future has its challenges. Significant challenges, including political resistance, economic interests, and social inequality, can impede climate action. Finding common ground and fostering collaboration across sectors will be crucial in overcoming these obstacles.

Universities and Research Institutions:

Universities and research institutions play a critical role in developing innovative solutions to climate change. Through rigorous research and development, these institutions can create groundbreaking technologies and strategies to mitigate climate impacts and accelerate the transition to a low-carbon future.

Furthermore, universities educate future leaders responsible for shaping climate policy and implementing sustainable solutions.

The Power of Public Engagement and Advocacy

Public engagement and advocacy are potent tools for shaping climate policy. Citizens who demand political action can significantly influence the policy landscape. Here is how different groups can contribute:

Individual Action and Advocacy: Individuals can amplify their voices through activism, voting for climate-conscious candidates, and making sustainable lifestyle choices.

Universities and Academic Institutions: Universities can promote public engagement through educational outreach programs and by disseminating research findings. They can also educate future leaders who are equipped to address climate challenges.

Movements and Campaigns: Grassroots and advocacy campaigns are crucial in raising public awareness, pressuring policymakers, and influencing political agendas. These groups can mobilize public support for climate action and hold governments accountable for their commitments.

Citizens can ensure that environmental sustainability remains a top priority for policymakers by actively participating in the political process and advocating for climate action. Voting for candidates with robust environmental platforms is critical to translating public concern into concrete policy change.

Through collaboration, innovation, and an unwavering commitment to a sustainable future, we can overcome the challenges of climate change and build a cleaner, healthier planet for future generations.

CHAPTER 9
Uniting Generations For The Planet

"Alone, we can do so little; together, we can do so much."

— Helen Keller

Climate change is a challenge that transcends generations. This chapter echoes Helen Keller's powerful message: "Alone we can do so little; together we can do so much." Collaboration is critical in the fight for a sustainable future. This chapter will explore how people of all ages, from Millennials to Baby Boomers and Generation Z, can work together to address climate change effectively. We will examine each generation's unique strengths and perspectives and explore how collective action can become a powerful force for positive change.

Bridging the Generational Divide:

Climate change is a challenge that demands a united front. To effectively address this global crisis, we need collaboration across generations. This chapter explores how Millennials, Gen Xers, and Baby Boomers can bridge the gap and work together for a sustainable future.

Synergy Through Shared Knowledge and Technology:

Millennials: Respect the wisdom and experience of older generations. At the same time, Millennials can help Baby Boomers adapt to new technologies and utilize these tools to advance climate action.

Baby Boomers: Embrace the technological advancements championed by Millennials. Share your wealth of knowledge and experience to guide younger generations in the fight against climate change.

Gen X: Gen X is a vital bridge between Millennials and Baby Boomers. Gen X can play a crucial role in fostering collaboration by learning from both generations and leveraging their unique perspectives.

Building a Collaborative Movement:

Shared Vision and Values: Developing a common vision for a sustainable future and establishing shared values around environmental responsibility is essential.

Setting Common Goals: Working together to set clear, achievable goals for climate action will provide a roadmap for collective progress.

Collaborative Strategies: We can maximize our collective impact by collaborating on strategies to achieve these goals.

Fostering Trust: Building trust among all stakeholders – governments, businesses, and individuals – is critical for successful collaboration.

Through mutual respect, knowledge sharing, and a commitment to shared goals, we can bridge the generational gap and create a powerful movement for climate action.

Understanding Diverse Perspectives: A World of Solutions

Climate change is a global challenge, but it impacts people differently. To create effective solutions, we must acknowledge the varied circumstances people face worldwide. Here is why understanding diverse perspectives is crucial:

Considering Global Disparities: Developed and developing nations have vastly different experiences with climate change. Developed countries have historically contributed more greenhouse gas emissions, while developing nations often face the brunt of climate impacts despite lower emissions. Acknowledging these disparities is essential for crafting equitable solutions.

Moving Beyond a One-Size-Fits-All Approach: A "one-size-fits-all" approach to climate action will not work. Strategies must be tailored to address different communities' unique needs and vulnerabilities. For instance, climate adaptation strategies for coastal communities facing rising sea levels will differ from those needed for inland agricultural communities facing extreme weather events.

Crafting Solutions for Diverse Communities: Effective climate action requires culturally appropriate, economically viable, and socially just solutions for all communities. This may involve involving local communities in decision-making and ensuring climate solutions consider their needs and priorities.

Empathy and Understanding: Building a genuinely collaborative approach to climate action requires compassion and understanding. By recognizing the diverse perspectives and challenges different communities face, we can foster collaboration and develop solutions that benefit everyone.

By committing to understanding diverse perspectives and tailoring solutions to specific needs, we can take a more inclusive and practical approach to climate action.

IV. Stories of Collaboration: Bridging the Gap, Building a Future

Climate action thrives on collaboration, and stories of successful partnerships offer hope and inspiration. Here, we will explore examples of how different generations are working together to create a more sustainable future:

A. Intergenerational Mentorship Program

The Sacred Valley of Peru is rich in wildlife and heritage, tucked away amidst the majestic Andes peaks. Here, the wisdom of the elders is vital, having nurtured the land for generations using Inca agricultural methods. The younger generation seeks answers to the changing climate because they are full of tech-savvy inventiveness and a strong sense of sustainability.

In the Sacred Valley, an innovative intergenerational mentorship program was started this summer. Young people full of energy are coupled with seasoned farmers whose faces are etched with the tales of the land. The elders impart their tried-and-true wisdom on crop rotation and soil conservation, two traditional agricultural practices encouraging peace with the environment. The youth, in turn, provide cutting-edge concepts like hydroponics and sustainable energy options like solar-powered irrigation systems.

A beautiful connection of respect and common purpose blossoms as they collaborate on eco-projects and community gardens. Older adults believe their heritage is being upheld and their wisdom is honored. The young people get the courage to become environmental stewards with a thorough awareness of their land as they realize the power of their knowledge paired with traditional customs.

Actionable Steps:

Individuals: Consider volunteering your time or skills to initiate or participate in intergenerational mentorship programs in your community. If you possess knowledge about sustainable practices, share it with younger generations. Conversely, if you are young and tech-savvy, reach out to elders to learn from their wisdom and experience.

Organizations: Foster collaboration between generations by organizing workshops, knowledge-sharing sessions, and community gardens where people of all ages can learn from each other.

This heartwarming story from the Sacred Valley demonstrates the power of collaboration in bridging the generational gap and fostering a culture of environmental stewardship. By learning from each other and working together, communities can create a more sustainable future.

However, the impact of climate change is not merely a statistic; it is a lived experience. Here is a personal story that highlights the reality of climate change:

As a child in the United Kingdom, I recall snow piled high against the telegraph poles. Winters meant snowstorms and even helicopter deliveries of supplies. Now, a foot of snow seems like a luxury. Summers are scorching, forcing even my dad, in his nineties, to invest in a portable air conditioning unit. The weather patterns have undeniably shifted dramatically.

B. Global Virtual Summit:

Now, turn your attention to a virtual summit brimming with action. Global participants come together in this lively virtual arena to preserve the environment for the next generations. The diverse array of panels and workshops offering a wide range of perspectives is what makes the summit buzz. Expert environmentalists impart their knowledge to youthful entrepreneurs by demonstrating state-of-the-art eco-technologies. In conversations about sustainable development, gender equality advocates pave a new path by ensuring all perspectives are heard.

During a particularly noteworthy session, an Indigenous elder from Kenya's Maasai Mara National Reserve delivers ancient teachings on living in harmony with nature. His voice is rich with traditions dating back millennia. Next to him, a youthful scientist from Kenya, her eyes bright with optimism, presents ground-breaking findings on preserving biodiversity in the area. Attendees actively participate in lively discussions, share successful local initiatives, and create connections beyond regional bounds. By the end of the summit, people's hearts were full of motivation, their minds were overflowing with creative solutions, and their determination to take on global challenges had become steadfast.

Actionable Steps:

Individuals: Seek out and participate in global virtual summits or online conferences focused on climate action. These events offer a wealth of information, connect you with a worldwide network of passionate individuals, and expose you to diverse perspectives on sustainability. Many summits offer live streams or recordings, making them accessible from anywhere worldwide.

Organizations: Consider hosting or co-hosting a virtual summit on a specific climate issue relevant to your region. This can be a powerful way to spark dialogue, showcase local initiatives, and foster collaboration on a global scale.

This virtual summit story exemplifies the power of technology to connect people across borders and ignite a collective spirit of environmental action. By participating in these online events, individuals can become part of a global village working toward a more sustainable future.

C. Community Gardens and Eco-Projects

In a busy Kampala, Uganda neighborhood, the hum of positive transformation blends with the rhythmic sounds of city life. People of all ages and ethnicities are joining forces to revitalize abandoned lands and turn them into thriving community gardens.

With hands seasoned by decades of land cultivation, older adults impart their green thumb knowledge to enthusiastic kids. They carefully impart their knowledge to ensure the following generation is taught the delicate art of planting seeds and caring for crops. Full of imagination, teenagers create clever irrigation systems out of recycled materials, showcasing the value of creativity in environments with limited resources.

The foundation of this endeavor is cooperation. Residents work together to start composting programs, converting garbage into valuable fertilizer. They host workshops on sustainable living to educate and encourage one another to adopt sustainable practices. Gender and generational ties grow as they work hand in hand. Sharing their knowledge gives older people a fresh feeling of purpose and offers young people helpful life skills and a strong bond with nature.

Harvest festivals turn into happy occasions to celebrate the bounty that their combined efforts have generated. Because of these gardens' success, other neighborhoods are motivated to follow suit and bring positive change across the city.

Actionable Steps:

Individuals: Identify vacant or underutilized land in your community that could be transformed into a community garden. Contact local authorities, environmental organizations, or neighborhood groups to gauge interest and collaborate on a project proposal. Offer your skills and volunteer your time to help establish and maintain the garden.

Organizations: Consider providing resources or expertise to support community garden initiatives. This could involve offering workshops on sustainable gardening practices, donating seeds or tools, or helping to connect community members with relevant funding opportunities.

This story from Kampala exemplifies the power of community gardens to foster collaboration, environmental responsibility, and a sense of shared purpose. By taking the initiative and working together, individuals can cultivate change and create pockets of sustainability within their communities.

D. Youth-Led Environmental Campaigns

India is witnessing a surge in youth activism driven by a fierce desire for gender equality and environmental justice. Young leaders are central in the struggle for a sustainable future in thriving cities and small rural communities.

These youthful campaigners think that more than conventional approaches are required. They plan colorful marches that wind through urban neighborhoods as their calls for action are echoed by their cries. Vital, visually striking pieces of art made from repurposed materials serve as poignant reminders of the environmental problems they face. Their preferred weapon is social media, where they use well-planned campaigns to spread the word about essential issues like climate change and women's critical role in achieving sustainability.

A particularly effective initiative is a set of online workshops in which young Indian women share their tales of inventiveness and resiliency. These workshops allow young women to demonstrate their leadership and knowledge in typically male-dominated subjects like environmental policy and renewable energy.

They continued after that. The initiative aggressively raises the voices of Native American women, spearheading grassroots campaigns for climate adaptation and sustainable land management. Together with their male allies, they push for inclusive solutions that cater to every community's unique requirements.

This initiative is making a significant impact as it gathers steam. A new generation of activists is rising to join the fight, motivated by the movement's intersectional approach to environmentalism, and policymakers are taking note.

Actionable Steps:

Individuals: Are you passionate about climate action and social justice? Get involved with existing youth-led environmental organizations in your community. If such organizations don't exist, consider starting one yourself. Social media is a powerful tool; use your platform to raise awareness about environmental issues and advocate for change.

Organizations: Provide support and resources to youth-led environmental campaigns. This could involve offering mentorship, workshop space, or assistance amplifying their message through your networks.

This story from India demonstrates the power of youth activism to spark change and inspire a more just and sustainable future. By harnessing their passion, creativity, and technological prowess, young people are at the forefront of building a better tomorrow.

E. Digital Storytelling Platforms

Envision a lively digital platform where individuals from diverse backgrounds in the Philippines convene to exchange their environmental experiences. With a voice full of pride for her commitment to a sustainable future, a retired engineer with a face engraved with experience's wisdom looks back on a lifetime of developing energy-efficient buildings.

Through a video blog, a young climate activist across the digital divide chronicles her journey from local protests to global advocacy. Her fervent remarks brim with the enthusiasm of a young person driven to change the world.

Beyond national boundaries and linguistic barriers, these narratives promote compassion and harmony within a worldwide society committed to environmental preservation. Different viewpoints come to light, creating a complex web of experiences. Elders' accounts of how industrialization affected their communities are essential historical records. Filled with creative ideas, young entrepreneurs present their solutions to the plastic pollution problem, encouraging others to follow suit.

Collaborative efforts flourish in comment sections and stimulating online debates. With the same goal, viewers of all ages and genders unite to establish bonds. Through this platform, they hope to create a future in which every narrative serves as a catalyst for change and every voice helps to create a better planet.

Actionable Steps:

Individuals: Seek out digital storytelling platforms focused on environmental issues. Share your own story about your environmental journey, big or small. It could be anything from transitioning to a plant-based diet to starting a community composting project. Your story has the power to inspire others and spark change.

Organizations: Consider developing or supporting digital storytelling initiatives in your community. This could involve creating an online platform, hosting storytelling workshops, or partnering with existing platforms to amplify local voices.

This story from the Philippines demonstrates the power of digital storytelling to connect people, share knowledge, and inspire collective action. By sharing our experiences and learning from each other, we can build a more sustainable future, one story at a time.

F. Cultural Exchange Programs

Imagine an innovative exchange program that links young people in urban São Paulo, Brazil, with native populations living deep within the Amazon rainforest. These young urbanites, used to the city's fast-paced lifestyle, are thrust into a society prosperous in traditional knowledge and a deep reverence for the natural world.

The Yanomami tribe's elders gently impart ancient farming methods passed down through the ages. Their features are engraved with the memories of decades of living in harmony with the rainforest. For the visiting kids, these tried-and-true methods—such as crop rotation and sustainable harvesting techniques—offer insightful lessons in sustainable living.

Conversely, the youthful, tech-savvy residents of São Paulo offer another set of abilities. By imparting their expertise on digital technologies for environmental activism and monitoring, they enable the Yanomami population to use technology to safeguard their territory. Drones with high-definition cameras can record illicit deforestation, and social media can spread the word about the value of protecting the Amazon rainforest worldwide.

These cultural exchange initiatives foster the sharing of knowledge and understanding across different cultures. Common ideals that cut across cultural divides, such as respect for the environment and gender equality, serve as the cornerstone for enduring connections between these disparate groups.

After returning home, the participants take on the role of change agents. The Yanomami use newly acquired technologies to fight for their land rights worldwide, while the urban youngsters incorporate the sustainable habits they learned from them into regional environmental efforts. This exchange program is a beautiful example of how cooperation can create a more sustainable future for everybody.

Actionable Steps:

Individuals: Research cultural exchange programs focused on environmental issues. If you find a program that aligns with your interests, consider applying to participate. These programs offer a unique opportunity to learn from different cultures and contribute to global environmental efforts.

Organizations: Advocate for creating or expanding cultural exchange programs focused on environmental sustainability. This could involve partnering with indigenous communities, educational institutions, or ecological NGOs to develop meaningful exchange opportunities.

Cultural exchange programs can foster cross-cultural understanding and collaboration, accelerating progress toward a sustainable future.

G. Art and Performance Festivals

Imagine the streets of Yogyakarta, Indonesia, teeming with a bright art festival. Recycled materials sculptures, such as used plastic bottles turned into colorful flowers and scrap metal welded into fantasy creatures, are boldly displayed alongside enthralling performances that honor the tenacity and beauty of nature.

Artists of many ages and origins are featured prominently. With years of experience, elders use their hands to weave complex textiles that pay tribute to the abundant biodiversity of their island home. Young dancers create performances motivated by the urgency of climate advocacy, expressing genuine emotion via their movements.

The festival offers an interactive experience as well as a visual extravaganza. Engaging artworks makes audiences consider environmental issues and the relationship between gender dynamics and ecological well-being. Poets take the platform, weaving dreams of a sustainable future through their poems, while panel discussions ignite spirited discussions about how policies affect communities of color.

Language barriers are broken down as participants create connections through spoken word, dance, music, and visual art—a shared commitment to social justice and planetary health anchors this vivid artistic mosaic. The festival makes a lasting impression, motivating participants to turn their creative experiences into tangible steps toward a future with greater sustainability.

Actionable Steps:

Individuals: Look for art and performance festivals with an environmental focus in your community. Get involved! Attend the festival, participate in workshops, or submit artwork reflecting your ecological concerns. Art can be a powerful tool for raising awareness and inspiring action.

Organizations: Consider organizing or supporting an art and performance festival focused on environmental issues. This could involve partnering with local artists, environmental NGOs, or community centers to create a platform for artistic expression and dialogue around sustainability.

H. Online Learning Platforms

Picture a dynamic virtual education system throughout India's vast and varied terrain. This platform provides an extensive range of courses on sustainable development to give students of all ages and backgrounds the information and abilities they need to build a more environmentally friendly future.

An elderly professor from the Indian Institute of Technology Delhi, whose face is engraved with the wisdom of a lifetime studying climate science, provides insightful observations from decades of work on climate resilience. Students from all around the nation find his lectures insightful and passionate.

In addition to this seasoned viewpoint, there is a youthful businessperson from a remote Maharashtra village. Her online environmentally responsible business practices course is jam-packed with helpful knowledge and real-world applications. Her contagious energy and creative attitude motivate students to integrate sustainability ideas into entrepreneurial action.

In virtual classrooms, learners from all over India actively participate, creating a vibrant learning atmosphere. Talks take place anywhere in the world, utilizing the varied perspectives and knowledge of students from big cities to small towns and everywhere in between. Through project-based learning, students address real-world issues such as deforestation in the Western Ghats and water scarcity in dry regions. Collaboration is essential to this process. They cultivate a mindset of cooperative problem-solving through exchanging accomplishments and setbacks, enabling mutual learning.

This online platform helps a committed global learning community come together to promote inclusive growth and environmental sustainability.

Actionable Steps:

Individuals: Explore online learning platforms offering courses on environmental sustainability. Numerous platforms are available, many of which offer free or low-cost classes. Enroll in a course that aligns with your interests and empowers you to contribute to a sustainable future.

Organizations: Consider developing or supporting online learning initiatives focused on environmental education. This could involve creating your courses, partnering with existing platforms, or providing scholarships to learners from underprivileged communities.

I. Corporate and NGO Partnerships in Pakistan

In Pakistan, a multinational corporation known for its commitment to clean energy joined forces with a local NGO dedicated to empowering women entrepreneurs. This innovative partnership focused on creating a brighter future powered by renewable energy and female leadership.

Together, they launched a series of training programs designed to equip women with the skills they need to thrive in the green economy. These programs provide comprehensive training in solar panel installation, maintenance, and sustainable business practices. Women entrepreneurs gain the knowledge and confidence to establish their businesses, bringing clean energy solutions to communities across Pakistan.

The corporation's resources go beyond technical training. Scholarships are offered to deserving young women, paving the way for them to pursue higher education in environmental science and engineering. These future leaders will be at the forefront of developing innovative solutions to Pakistan's climate challenges.

Mentorship programs bridge the gap between generations and genders. Aspiring female leaders are paired with seasoned professionals in the renewable energy sector. Through mentorship, valuable knowledge is transferred, and role models emerge, inspiring a new wave of women to enter this critical field.

This public-private partnership is not just about skills development but advocacy. Together, the corporation and NGO champion gender-responsive climate policies that ensure women have a seat at the table and benefit from the transition to a green economy.

They advocate for clean technologies that empower local communities and promote sustainable development across Pakistan.

This story exemplifies the power of collaboration between corporations and NGOs. By working together, they can unlock the potential of renewable energy, empower women, and build a more sustainable future for all.

Actionable Steps:

Individuals: Research and support corporations with strong environmental and social responsibility practices. Consider investing in companies that partner with NGOs on sustainability initiatives. Get involved with NGOs working on climate issues in your community. Offer your skills and volunteer your time to support their projects.

Organizations: Explore partnership opportunities with corporations that align with your mission. Consider co-creating training programs, mentorship initiatives, or advocacy campaigns focused on environmental sustainability.

J. Policy Advocacy and Grassroots Movements in Kenya

A powerful movement is taking root in the heart of Kenya's Rift Valley. Led by wise Maasai elders, steeped in the traditions of their ancestors and deeply connected to the land, a grassroots movement is fighting for the rights of indigenous communities and environmental justice.

These elders, their voices seasoned with the wisdom of generations living in harmony with nature, are not alone in their fight. They have joined forces with passionate youth activists, forming an intergenerational alliance fueled by a shared vision for a sustainable future.

Together, they have embarked on a determined campaign for policies that protect the rich biodiversity of their ancestral lands. They advocate for sustainable land management practices and promote gender equality in resource management decisions, ensuring that women have a rightful voice in shaping the future of their communities.

Their fight does not stop at local borders. They actively lobby policymakers at both the national and international levels. Armed with compelling stories of resilience in the face of environmental degradation and unwavering resistance against land grabs, they capture hearts and minds.

They influence legislative agendas through inclusive dialogue and community organizing, securing crucial support for their cause. Victories, big or small, fuel their movement's growth. With each win, marginalized communities can reclaim their voices and stewardship over their ancestral lands.

Actionable Steps:

Individuals: Stay informed about environmental and social justice issues in your community. Get involved with local organizations working on these issues. Raise awareness through social media, community events, or conversations with friends and family. Contact your elected officials to voice your support for policies promoting environmental protection and indigenous communities' rights.

Organizations: Support grassroots movements in your region working on environmental and social justice issues. This could involve offering capacity-building workshops, fundraising support, or assisting with advocacy efforts.

V. Conclusion:

The narratives we have read and discussed present a compelling picture of cooperation, a cross-cultural patchwork of respect, solidarity, and a common goal. We witness young and older people exchanging wisdom, technology bridging gaps, and art igniting a shared enthusiasm for a sustainable future.

The timeless proverb "Alone we can do so little; together we can do so much" by Helen Keller reverberates throughout these stories. Every narrative demonstrates the transforming potential of teamwork. We can create a more equitable and sustainable society by sharing knowledge and skills, learning from one another, and motivating one another to take action.

Our problems are multifaceted and interrelated, including social injustice, environmental degradation, and climate change. However, as these tales show, we are not helpless. When we join behind a shared objective, we can use humanity's collective wisdom, energy, and creativity to bring about significant change. Let us draw inspiration from these tales. Equipped with understanding, empathy, and a dedication to working together, let us set out to create a more promising future for everybody.

Conclusion

Growing up in Staffordshire, surrounded by the lush green landscapes that shaped my childhood, I learned firsthand the delicate balance between human activity and protecting the environment. These early experiences have given me a unique blend of traditional wisdom and practical sustainability knowledge.

My lifelong passion has been exploring environmental challenges, and this book is a testimony to my desire to share information that inspires action. I strive to keep the dialogue alive through personal initiatives, community involvement, and more considerable advocacy efforts focused on sustainable agriculture, renewable energy, and environmental conservation.

I hope my work inspired you. It carries a powerful message that can positively change your life and community. Together, let us wholeheartedly embrace the principles of sustainability and conservation as the very foundation for a healthy and resilient future for our planet. By sharing my journey and lessons, I aim to inspire others to reflect on their relationship with nature and discover their potential as change-makers.

The challenge posed by the climate crisis has no easy answers. It requires a collective, concerted, and coordinated effort from each generation. Every generation has played a crucial role in addressing climate change, each bringing something unique. In this book, I identify the wisdom and experience of the Baby Boomers, the bridge-building qualities of Generation X, the innovation and activism of Millennials, and the insistence on accountability from Generation Z. It is through collaboration and embracing diversity that we can strive for a better and more sustainable future.

Let us draw inspiration from the countless individuals and communities worldwide working tirelessly to save our planet. Their stories show how people can be adamant when they need to be and what we can achieve when we come together for positive change. Let us create a world where humans and nature can coexist harmoniously.

Millennials have taken center stage on the world platform. They deeply understand the climate crisis and have been leading the way in sustainable movements for decades. This generation approaches activism through a unique lens, with a solid focus on solutions and a fearless commitment to speaking out against injustice.

Compared to previous generations, millennials are unsatisfied with mere information; they demand action. They are willing to engage in civil disobedience when necessary and utilize digital media as a platform to amplify their voices. This focus on environmental education by formal institutions may pave the way for a profoundly ethical generation.

This shift in mindset is evident in millennials' choices and lifestyles. From embracing vegan diets to making conscious decisions that result in less waste, this generation is acutely aware of the collective impact of their actions. They recognize the potential for change to take shape through community-driven sustainability efforts.

Social media platforms like Twitter (X), Instagram, and TikTok have played a vital role in sharing information, facilitating conversations, and mobilizing people. Millennials, in particular, can leverage these platforms to create viral content that raises awareness, promotes correctness, and holds leaders accountable.

Millennials redefine the essence of environmental activism through their adept use of technology and proactive approach. Their passion, innovation, and unwavering commitment to a greener future inspire hope for future generations.

The fight for a livable future transcends generational boundaries. To be effective, it requires bridging the gaps between generations and fostering intergenerational collaboration among Millennials, Gen Xers, and Baby Boomers. By coming together under one umbrella, we can combine our generational strengths, pool resources, and galvanize collective power to drive meaningful change.

Drawing on their experience and financial capabilities, baby boomers can substantially impact the battle against climate change. They can invest in renewable energy, support climate-friendly policies, and mentor the younger generation. With their unparalleled understanding of technological and Economic changes, Gen Xers can act as a bridge between generations, advocating for policy shifts, driving sustainable business practices, and providing mentorship to the next wave of environmental leaders.

Millennials have emerged as a formidable force for social change through their digital prowess and innovation. Their adept use of social media platforms to organize people has raised awareness and fostered activism. Generation Z, building upon Millennials' work, pushes for greater accountability and demands systemic change.

Encouraging and appreciating each generation's contributions and fostering collaboration can create a powerful movement for change. Let us join forces to build a world where humans and nature thrive harmoniously.

As you reach the end of this book, please take a moment to reflect on the compelling stories and invaluable lessons shared within its pages. Each chapter illuminates the importance of collaboration, innovation, and determination in the face of climate change.

Now, it is your turn to become the change you wish to see in the world, regardless of whether you are a Baby Boomer, a wise Gen Xer, an idealistic Millennial, or a passionate Gen Zer. You have a unique role to play in this effort. Embrace the opportunity to make a difference.

Lead a sustainable life daily, inspire change within your community, and motivate others to join the cause. Together, let us create a society that embraces renewable energy and minimizes waste, respecting our planet's delicate balance.

Make no mistake: if we continue disregarding climate change, the consequences will be devastating. We face the risk of severe droughts, widespread famine, and the displacement of millions of people from their homes. However, we have the power to change this bleak future.

While we expect governments to take more decisive action, implementing stringent regulations, investing in renewable energy sources, and participating actively in international agreements to reduce the greenhouse effect, the responsibility does not solely lie with politicians. Businesses, too, have a crucial role to play. They should adopt ecologically sound practices, reduce waste, and invest in research for green technologies that can contribute to a more sustainable future.

Climate change knows no borders. It is a global challenge that requires countries worldwide to work together, share knowledge and resources, and develop new technologies.

As an individual, you might wonder, "What can I do?" The truth is, your actions matter. You become part of the solution whenever you consciously live sustainably or advocate for climate-friendly policies. Moreover, when millions of us join in, real change happens.

The enormity of the challenge can be overwhelming, and it may often feel like progress could be faster. However, this book is not about dwelling on the problems; it is about highlighting solutions and showing how we can contribute to building a more sustainable and resilient future.

Throughout the book, I have shared inspiring stories of people coming together and translating innovative ideas into tangible products and solutions. These stories serve as a roadmap for how we can all make a difference.

Remember, change happens step by step. Whether you support local businesses that prioritize sustainability, make environmentally conscious product choices, or reduce your carbon footprint, every action you take makes a positive impact. Engage in community initiatives, volunteer for green causes, and participate in local decision-making regarding climate issues to the best of your ability.

Never underestimate the power of one individual or you to create change. Let your voice be heard, your actions be impactful, and your spirit inspire others. Together, we can make a future where we get along with nature instead of fighting it.

Share your experiences and insights through book reviews or online platforms. Raising awareness and inspiring others is a powerful way to make a difference. Let the ripple effect of positive change spread to every corner of the earth.

As we conclude this journey together, I invite you to leave a review of the book. Your feedback and insights are invaluable, not only in shaping future works but also in inspiring others to join our cause. I eagerly look forward to learning from your thoughts and experiences as we continue our quest to address these vital themes.

By harnessing our collective wisdom, energy, and creativity, we can tackle the critical problems of our time. Remember that even the most minor steps, taken consistently, can lead to significant change. Each action has far-reaching effects, whether installing a solar system, reducing plastic waste, practicing aggressive recycling, opting for paper bags, conserving water, or using energy-efficient appliances. It extends beyond your immediate surroundings to impact communities, states, countries, and the world.

Let us also recognize the wisdom and knowledge that elders possess. Instead of engaging in blame games, let us create space for discussion and learn from their insights. This book emphasizes the importance of intergenerational communication and values the vast experiences of baby boomers and elders.

Allow me to share a personal anecdote that perfectly illustrates this point. A few weeks ago, despite living in a cloudy UK, my 90-year-old father recently decided to supplement his long-standing solar system with a vertical wind turbine. He recognized the limitations of solar power in the British climate and turned to this innovative technology. Compact and efficient, vertical wind turbines can harness energy from any direction, making them an appealing option for supplementing renewable energy production.

His adaptability and openness to new technology inspire me, especially at his age.

This story emphasizes the wealth of knowledge we can gain from older generations. They have witnessed tremendous changes and often offer unique perspectives on sustainability. We must be receptive to their ideas, engage in meaningful dialogues, and provide guidance on emerging technologies. Of course, this does not mean disregarding their experiences and the proven practices of the past. As we forge ahead, let us continue to inspire and take action.

Commit to doing something small that makes a difference each day, no matter how seemingly insignificant. Please share your journey to inspire and motivate others to join our cause. Remember, every individual action matters, and together, we can create a powerful wave of positive change that shapes a sustainable future for generations to come.

Your involvement in this movement is essential. Only through united efforts, the sharing of knowledge, and continuous action can we address the many environmental challenges we face. This is our opportunity to leave a lasting impact on our world. As we reach the final page of this book, it serves as a reminder that it is merely the beginning of our journey toward a more sustainable future. The destiny of our planet lies in our hands. Let us work together to make it a bright one.

References

Aarhus University. (2023). International collaboration on food waste. https://international.au.dk/

Africa Progress Group. (2018). Now is a time for action, affirms Kofi Annan in the final Africa Progress Panel report. Retrieved from [africaprogressgroup.org] (http://africaprogressgroup.org/index.php/2018/03/02/now-is-a-time-for-action-affirms-kofi-annan-in-final-africa-progress-panel-report/)

Amazon Conservation Association. (2023). Climate Change Threats to the Amazon.

Amazon Sustainability. (2023, January 17). The Earth Fund. https://sustainability.aboutamazon.com/

American Forests. (2023, January 11). How much plastic waste is produced each year? https://www.statista.com/study/65164/plastic-waste-worldwide/

Apple. (2022, April 8). Environment. https://www.apple.com/environment/

Azeez, J. A. (2023). Climate change risks and adaptation strategies for smallholder farmers in developing countries. Journal of Environmental and Agricultural Sustainability in Developing Countries, 12(2), 1–12. https://ipad.fas.usda.gov/highlights/2023/05/NorthwestAfrica/index.pdf

Barnes, A. (2022, March 25). The Hill. The Hill. https://thehill.com/

BASF. (2023). ChemCycling™: Chemical recycling of plastics. https://www.basf.com/

BBC News. (2023, March 14). Electric cremation: A greener way to say goodbye. https://www.bbc.com/future/article/20230821-how-to-plan-a-sustainable-funeral

BC Hydro. (2022, March 22). Report: Gen X is the 'bridge generation' between baby boomers and millennials regarding climate change action. https://www.bchydro.com/energy-in-bc/operations/generation.html

Betts, A. (2020). The future of forced displacement: A conceptual framework. Refugee Survey Quarterly, 39(4), 381–400. [doi:10.1093/rsq/hqaa030]

Bill & Melinda Gates Foundation. (2024, April 10). Water, Sanitation & Hygiene. https://www.gatesfoundation.org/our-work/programs/global-growth-and-opportunity/water-sanitation-and-hygiene

BMU Bundesministerium für Umwelt, Naturschutz und Reaktorsicherheit. (2021, January 19). The German Renewable Energy Sources Act (EEG). https://www.iea.org/policies/5734-2014-amendment-of-the-renewable-energy-sources-act-eeg-2014

Boykoff, M. T., & Yulsman, K. (2013). Political discourse on climate change: Political polarization and scientific uncertainty. Environmental Communication, 7(1), 95-112. https://onlinelibrary.wiley.com/page/journal/19401582/homepage/

productinformation.html

Brookings. (2023). India's Climate Change Policy in a Modi Government. Retrieved from https://www.brookings.edu/articles/indias-climate-change-policy-in-a-modi-government/

Byjus. (2023). Importance of renewable energy - types of renewable energy and resources. https://byjus.com/physics/renewable-energy/

Carrico, A., Zell, E., & Cohn, A. (2021). Prioritizing the near and dear? How poverty shapes environmental risk perceptions and mitigation behaviors. Nature Climate Change, 11(4), 318-323. https://www.nature.com/nclimate/articles

Carrington, D. (2021, April 22). Climate activists target Barclays Bank in London protests. The Guardian. https://www.theguardian.com/business/2023/dec/27/uk-students-launch-barclays-career-boycott-over-banks-climate-policies

Central Pollution Control Board, India. (2023). Ambient Air Quality Monitoring. https://cpcb.nic.in/National-Air-Quality-Index/

Climate Reality Project. (n.d.). About Al Gore. https://www.climaterealityproject.org/

CNBC. (2023, March 1). Elon Musk. https://www.cnbc.com/2023/05/16/elon-musk-cnbc-interview-with-david-faber.html

Coastal zone management. (n.d.). UNEP - UN Environment Programme. https://www.unep.org/topics/ocean-seas-and-coasts/regional-seas-programme/coastal-zone-management

Consumer Ecology. (2022, June 23). *Beef Carbon Footprint & Environmental Impact - Consumer Ecology*. Consumer Ecology. https://consumerecology.com

Cornell University, College of Agriculture and Life Sciences. (2023, March 14). Crop rotation. https://cvp.cce.cornell.edu/submission.php?id=751

Crippa M., Guizzardi D., Banja M., Solazzo E., Muntean M., Schaaf E., Pagani F., Monforti-Ferrario F., Olivier, J.G.J., Quadrelli, R., Risquez Martin, A., Taghavi-Moharamli, P., Grassi, G., Rossi, S., Oom, D., Branco, A., San-Miguel, J., Vignati, E. CO2 emissions of all world countries – JRC/IEA/PBL 2022 Report, Publications Office of the European Union, Luxembourg, 2022, doi:10.2760/07904, JRC130363. https://edgar.jrc.ec.europa.eu/report_2022

Debate. The National Academies Press. https://doi.org/10.17226/21881

Deutsche Welle. (2022, November 21). Climate change is displacing millions in the Philippines. https://www.climate-refugees.org/perspectives/2023/5/8/philippines

El-Hinnawi, E. S. (1985). Environmental refugees. Natural Resources Forum, 9(1), 187–191.

Ellen MacArthur Foundation. (2023, January 18). A new textile economy: Redesigning fashion's future.

Enerkem. (2023). Waste-to-biofuels and renewable chemicals. https://enerkem.com/

Engel, S., Wunder, S., & Pagiola, S. (2008). Expanding the de facto forest conservation area in Costa Rica through payments for environmental services. IfPRI Discussion Paper, (00834).

Environmental Protection Agency. (2024, January 19). Clean Air Act Overview. https://www.epa.gov/clean-air-act-overview/clean-air-act-text

Environmental Protection Agency. (2021, January 19). Types of EPA Regulations. https://www.epa.gov/laws-regulations

Environmental Protection Agency. (2020, April 14). Sources of Greenhouse Gas Emissions.

European Commission. (2023, February 14). EU agriculture policy: Supporting farmers and enhancing rural areas. https://agriculture.ec.europa.eu/index_en

European Commission. (2023, May 20). The CAP: Greener, fairer, stronger. https://www.ec.europa.eu/info/food-farming-fisheries/key-policies/common-agricultural-policy_en

European Commission. (n.d.). The European Green Deal. Retrieved from https://ec.europa.eu/commission/presscorner/detail/en/ip_19_6691

European Commission. (2023, July 14). EU Emissions Trading System (EU ETS). https://climate.ec.europa.eu/eu-action/eu-emissions-trading-system-eu-ets_en

Eurostat. (2022, November 30). Waste statistics. https://ec.europa.eu/eurostat/statistics-explained/index.php/Waste_statistics

FAO. (2023, April 12). Climate change and food security. https://www.fao.org/3/i5188e/I5188E.pdf

Freelancers Bangladesh. (n.d.). Freelancers Bangladesh – Freelancing Development Ecosystem. https://freelancers.gov.bd/

Fridays for Future. (2023, March 3). History of Fridays for Future. https://fridaysforfuture.org/

Food and Agriculture Organization of the United Nations. (2023, February 15). Deforestation - The causes, consequences, and solutions.

Food and Agriculture Organization of the United Nations (FAO). (2021). Global Forest Resources Assessment 2020. https://www.fao.org/forest-resources-assessment

Food and Agriculture Organization of the United Nations [FAO]. (2023, May 12). Sustainable agriculture: Meeting the needs of the present without compromising the ability of future generations to meet their own needs. https://www.fao.org/sustainability/en/

Generation Next: Gen X overtake Baby Boomers as largest carbon emitters in B.C. (2022, October 21). https://www.bchydro.com/news/press_centre/news_releases/2022/report-gen-x-baby-boomers.html

Getz, C., & Pickering, D. (2023). Agri-tourism. https://iopscience.iop.org/article/10.1088/1755-1315/421/2/022058

Gilmour, E., Thorne, J., Lopez-Carmona, J., & Berry, P. (2021). Adapting livestock production systems to climate change. Animal Production Science, 61(8), 1189–1210. https://onlinelibrary.wiley.com/doi/full/10.1002/jcp.28411

Green Climate Fund. (n.d.). Green Climate Fund. https://www.greenclimate.fund/

Greenpeace International. (n.d.). Greenpeace International. https://www.greenpeace.org

Global CCS Institute. (2023, May 12). About CCS. https://www.globalccsinstitute.com/

Global Reporting Initiative. (n.d.). Global Reporting Initiative (GRI). https://www.globalreporting.org/

Guttikunda, N. S., Mittal, S., Garg, A., Sharma, S., Santra, A. K., & Chakrabarty, M. (2023). Status of crop residue burning in agricultural states of India and potential mitigation strategies. Environmental Science and Pollution Research, 30(8), 12225-12240. https://www.matec-conferences.org/articles/matecconf/pdf/2024/01/matecconf_sesam2023_00066.pdf

Google. (2022, December 9). Sustainability. https://sustainability.google/

Gore, A. (2006). An inconvenient truth [Documentary film]. Participant Productions.

Gore, A. (2007). Nobel Lecture. Nobel Prize Outreach AB. Retrieved June 25, 2024, from https://www.nobelprize.org/prizes/peace/2007/gore/lecture/

Hindustan Times. (2021, November 21). Varanasi crematoriums struggle for wood as deforestation rises. https://www.newsweek.com/india-reports-shortage-wood-needed-funeral-pyres-covid-deaths-surpass-234k-1589571

History Channel. (2024, February 27). The Gulf War. https://www.history.com/topics/middle-east/persian-gulf-war-video

IBM. (2020, September 22). Blockchain for a Smarter Waste Management Ecosystem. [IBM Blog] https://www.ibm.com/blog/revolutionizing-the-waste-supply-chain-blockchain-for-social-good/

IEA [International Energy Agency]. (2023, October 27). AI for climate action. https://www.iea.org/

Iles, A. (2020). Building resilience through farmer cooperatives in Africa. Journal of Rural Studies, 36(4), 1426–1440. https://www.elsevier.com/

Intergovernmental Panel on Climate Change. (2021). Climate Change 2021: The Physical Science Basis. (Masson-Delmotte, V., P. Zhai, A. Pirani, S. L. Connors, C. Péan, S. Berger, et al. (Eds.)). Cambridge University Press.

Institute for Transportation and Development Policy (ITDP). (2013). Transforming Transportation in Bogota.

https://itdp.org/2023/06/26/city-as-a-learning-lab-mobilize-itdp-bogota/

Internal Displacement Monitoring Centre (IDMC). (2023). Climate Change and Displacement. https://www.internal-displacement.org/sites/default/files/publications/documents/grid2021_idmc.pdf

International Monetary Fund. (2020, October 9). Fiscal Policy Response to COVID-19. https://www.imf.org/en/Topics/imf-and-covid19/Policy-Responses-to-COVID-19

International Institute for Sustainable Development (IISD). (2023, March 3). Kenya's mobile early warning system helps pastoralists adapt to climate change. https://www.adaptation-undp.org/somalia-and-undp-launch-new-us10-million-project-pastoralist-communities-access-scarce-water

Intergovernmental Panel on Climate Change (IPCC). (2021). Climate Change 2021: The Physical Science Basis.

International Renewable Energy Agency (IRENA). (2023). Renewable Capacity Statistics 2023. https://www.irena.org/Publications/2023/Mar/Renewable-capacity-statistics-2023

International & Executive Programs, UC Berkeley. (n.d.). Kofi Annan: A Legacy in Global Environmental Diplomacy. Retrieved from [iep.berkeley.edu](https://iep.berkeley.edu/people/kofi-annan)

International Renewable Energy Agency. (2023). China. https://www.irena.org/

International Renewable Energy Agency. (2023). Renewable Energy Statistics 2023. https://www.irena.org/

IRENA. (2023, March 2). Renewable energy targets and gaps. https://www.irena.org/

IRENA [International Renewable Energy Agency]. (2023, April 11). Battery storage. https://www.irena.org/

Kelly, C. P., Oppenheimer, M., & Pettorelli, N. (2015). Climate change, data gaps, and the challenge of building climate-informed adaptation strategies in fragile states. Environmental Research Letters, 10(12), 120004. [doi:10.1088/1748-9326/10/12/120004]

Knuth, D. E. (2008). The Art of Computer Programming (3rd ed.). Addison-Wesley Professional.

Labiak, M. (2024, June 12). European Union threatens China EVs with tariffs of up to 38%. https://www.bbc.com/news/articles/cd11ze1k9r0o

Land Trust Alliance. (2023, February 14). Leaving a legacy: Conservation easements and estate planning. https://landtrustalliance.org/take-action/give/ways-to-give/planned-giving

Lennox, S., & Sacks, D. (2014). Listicle: The new news? Journalism Studies, 15(7), 862–880. https://www.tandfonline.com/journals/rjos20

Lichtenstein, A. (2020, September 24). Xiye Bastida: The 17-year-old leading the fight for climate justice. Teen Vogue. https://www.teenvogue.com/video/watch/how-an-indigenous-teen-climate-activist-plans-to-change-the-world

Li, X., Ma, C., Lin, S., Wang, Z., Zhao, L., & Li, J. (2021). Climate change and risk perception among farmers in a mountainous region of Central China. Sustainability (Switzerland), 13(18), 10322. https://www.mdpi.com/2227-9059/10/10/2437

McCombs, M. E., & Shaw, D. L. (1972). The agenda-setting function of mass media. Public Opinion Quarterly, 36(2), 176–18. https://onlinelibrary.wiley.com/page/journal/19401582/homepage/productinformation.html

Marchetti, C., Canzler, W., & Schwanen, T. (2017). Transportation and the future of cities. In Handbook of transport economics (Vol. 5, pp. 1-39). Edward Elgar Publishing.

McNamara, K. E., & Connell, J. (2021). Disappearing paradise: The existential threat of climate change to Pacific island communities. Environmental Reviews, 29(1), 78–89. [doi:10.1139/er-2020-0080]

Meland, S. I. (2022, June 21). People over 60 are greenhouse gas emission bad guys. Norwegian SciTech News. https://www.earth.com/news/greenhouse-gas-emissions-reach-another-all-time-high/

Michigan State University, Integrated Pest Management Program. (2023, March 1). What is IPM? https://www.canr.msu.edu/ipm

Ministry of Economy, Trade and Industry (METI), Japan. (2023, March 23). Recycling Policies in Japan. https://japannews.yomiuri.co.jp/politics/politics-government/20231013-142824/

Ministry of New and Renewable Energy, Government of India. (2023). Narendra Modi's renewable energy policies. Retrieved from https://mnre.gov.in/

Ministry of New and Renewable Energy, Government of India. (2023). India's progress toward renewable energy. Retrieved from https://mnre.gov.in/

MNRE, Government of India. (2023, March). Solar Energy. https://mnre.gov.in/

Modi, N. (2023). Prime Minister's visit to UAE for World Climate Action Summit. Retrieved from https://www.narendramodi.in/prime-ministers-visit-to-uae-for-world-climate-action-summit

Moyo, D. (2006). Dead aid: Why is aid not working, and how is there a better way for Africa? Farrar, Straus and Giroux.

MrBeast Philanthropy. (n.d.). About Us. https://www.beastphilanthropy.org/

National Academies of Sciences, Engineering, and Medicine. (2016). The Science of Climate Change: A Frame for the

National Aeronautics and Space Administration. (n.d.). Climate Change Evidence: How Do We Know? NASA (.gov). Retrieved March 13, 2024, from https://climate.nasa.gov/evidence/

National Geographic. (2019, September 23). Greta Thunberg: A teenage climate activist inspiring millions. https://www.natgeokids.com/uk/kids-club/cool-kids/general-kids-club/greta-thunberg-facts/

National Oceanic and Atmospheric Administration. (2023). Climate Data: National Centers for Environmental Information (NCEI). National Oceanic and Atmospheric Administration. Retrieved March 13, 2024, from https://www.ncei.noaa.gov/

Natural Resources Canada. (2023, April 19). Sustainable Forest Management. https://natural-resources.canada.ca/our-natural-resources/forests/sustainable-forest-management/13183

Nicholls, R. J., Marinova, N., Lowe, J. A., Brown, S., Vellinga, P., de Young, R., & ... Weaver, A. J. (2011). Sea-level rise and its impact on coastal zones. Reviews of Geophysics, 49(3). https://news.agu.org/press-release/worst-impacts-of-sea-level-rise-will-hit-earlier-than-expected/

NRDC [Natural Resources Defense Council]. (2023, May 10). Renewable energy definition - sources, clean alternatives. https://www.nrdc.org/issues/renewable-energy

Norwegian Environment Agency. (2023, February 22). Forests. https://www.miljodirektoratet.no/

Plastic Energy. (2023). Turning plastic waste into valuable resources. https://plasticenergy.com/

Rahman, M. M., & Salehin, M. S. I. (2018). A review of the effects of climate change on human health in Bangladesh. Journal of Environmental and Public Health, 2018, 1479025. [doi:10.1155/2018/1479025]

Resomation Ltd. (2024). What is Resomation? https://resomation.com/wp-content/uploads/sites/10/200406-The-eco-alternative-Natural-water-cremation-explained-American-Cemetery-and-Cremation-article.pdf

Rogelj, J., Ludlow, L., Cowie, A., Fricko, O., Guenther, A., Höhne-Sparling, L., ... & Meinshausen, M. (2023). Accelerating the transition to a low-carbon economy through financial aid to developing countries. Environmental Science & Policy, 133, 146-153. https://www.sciencedirect.com/science/article/abs/pii/S2590332223005547

San Francisco Environment. (2023). Achieving zero waste by 2030. https://sfenvironment.org/

Satterthwaite, D. (2018). Building resilience against climate and disaster risks in coastal cities. Environment and Urbanization, 30(2), 681–697.

Schroeder, R. (2020, February 24). How social media is changing the face of climate activism. The Guardian. https://www.theguardian.com/environment/2023/may/24/climate-activism-is-doomed-to-fail

Schroeder, A. (2019). Rhetorical Situation Analysis on Greta Thunberg | HONS 110-07. Retrieved from https://blogs.cofc.edu

Sierra Club. (n.d.). Sierra Club. https://www.sierraclub.org/

Solar Energy Industries Association. (2024, April 10). Investment Tax Credit (ITC). https://www.seia.org/initiatives/tax-policy

Speech to the British Parliament - April 23, 2019. (2019, December 3). Archives of Women's Political Communication.

https://awpc.cattcenter.iastate.edu/2019/12/03/speech-to-the-british-parliament-april-23-2019/

Task Force on Climate-Related Financial Disclosures. (n.d.). Task Force on Climate-Related Financial Disclosures (TCFD). https://www.fsb-tcfd.org/

Tesla. (2024, April 20). Our Impact. https://www.tesla.com/impact

TheCivilEngineer.org. (2024, March 22). The world's largest renewable energy project is being developed in India. *TheCivilEngineer.org*. https://www.thecivilengineer.org/news/worlds-largest-renewable-energy-project-is-being-developed-in-india

The Elders. (2024). Climate Change. Retrieved from [theelders.org](https://theelders.org/climate-change)

The Fab City. (2023). Urban upcycling initiative. https://www.fabcity.org/

The Federal Ministry for Economic Affairs and Climate Action. (2023, May 17). Energiewende – The German energy transition. https://www.bmwk.de/Navigation/EN/Home/home.html

The Guardian. (2021, July 21). Pacific nations plan for a future underwater. https://www.theguardian.com/world/2020/aug/29/pacifics-fight-against-covid-19-hamstrung-by-lack-of-clean-water

The Guardian. (2015). Kofi Annan: We must challenge climate-change skeptics who deny the facts. Retrieved from [theguardian.com](https://www.theguardian.com/environment/20

15/may/03/kofi-annan-interview-climate-change-paris-summit-sceptics)

The Guardian. (2017). Macron hosts climate summit without Trump. Retrieved from https://www.theguardian.com/environment/2017/dec/12/macron-hosts-climate-summit-without-trump

The New Humanitarian. (2023, March 21). Climate refugees: They are already here. https://www.thenewhumanitarian.org/environment-and-disasters/climate-change

The World Bank. (2019). Groundswell: Preparing for Internal Climate Migration. https://openknowledge.worldbank.org/handle/10986/31473

The World Bank. (2023, January 19). Climate change and agriculture. https://documents1.worldbank.org/curated/en/676601468249642651/pdf/Looking-beyond-the-horizon-how-climate-change-impacts-and-adaptation-responses-will-reshape-agriculture-in-Eastern-Europe-and-Central-Asia.pdf

The World Bank. (2023, April 12). Climate change. https://www.worldbank.org/en/topic/climatechange

The World Bank. (2020). Bangladesh: Investing in Climate-Resilient Infrastructure.

TWI global. (2023, January 11). Renewable energy definition - sources, clean alternatives. https://www.iea.org/energy-system/renewables

UNFCC [United Nations Framework Convention on Climate Change]. (2023, May 18). Technology for climate action. https://unfccc.int/

UK Department for Environment, Food & Rural Affairs. (2023, February 3). Countryside stewardship: Get started. https://www.gov.uk/guidance/countryside-stewardship-get-funding-to-protect-and-improve-the-land-you-manage

UN [United Nations]. (2023, May 18). What is renewable energy? https://www.unep.org/topics/energy/renewable-energy/renewable-energy

United Nations High Commissioner for Refugees (UNHCR). (1951). Convention Relating to the Status of Refugees. https://www.unhcr.org/1951-refugee-convention.html

United Nations Convention to Combat Desertification (UNCCD). (2021). Global Land Outlook 2: One Planet, Healthy Land. https://www.unccd.int/resources/global-land-outlook/glo2

United Nations Framework Convention on Climate Change (UNFCCC). (2015). The Paris Agreement.

United Nations Framework Convention on Climate Change (UNFCCC). (2023). Climate Action for Individuals.

United States Department of Agriculture. (2023, April 11). Climate change. https://www.usda.gov/oce/energy-and-environment/climate

United Nations. (n.d.). Secretary-General Ban Ki-moon (2007-2016). Retrieved from https://www.un.org/sg/en/former-sg/ban-ki-moon.shtml

United Nations Environment Programme. (2020, December 11). Technology transfer for climate change mitigation

United Nations Framework Convention on Climate Change. (2015, December 12). Paris Agreement. https://unfccc.int/process-and-meetings/the-paris-agreement

World Agroforestry. (2023, May 24). What is agroforestry? https://apps.worldagroforestry.org/treedb/

Warner, K. (2014). A typology of climate change-induced environmental migration. International Journal of Climate Change Strategies and Management, 6(4), 475–489. [doi:10.1108/IJCCSM-02-2014-0023]

White House. (2016). President Obama: The United States formally enters the Paris Agreement. Retrieved from https://obamawhitehouse.archives.gov/the-press-office/2016/09/03/president-obama-united-states-formally-enters-paris-agreement

White House. (2016). President Obama marks a historic moment in our global efforts to combat climate change. Retrieved from https://obamawhitehouse.archives.gov/the-press-office/2016/10/05/president-obama-marks-historic-moment-our-global-efforts-combat-climate

White House. (2016). U.S. leadership and the historic Paris Agreement to combat climate change. Retrieved from https://obamawhitehouse.archives.gov/blog/2015/12/12/us-leadership-and-historic-paris-agreement-combat-climate-change

Wikipedia. (2024, April 10). Sunita Narain. https://en.wikipedia.org/wiki/Sunita_Narain

World Bank. (2023). Climate Change and Migration. https://www.worldbank.org/en/news/press-release/2021/09/13/climate-change-could-force-216-million-people-to-migrate-within-their-own-countries-by-2050

World Bank. (2023, January 19). Private sector investment in agriculture: Is it crowding out public investment? https://documents1.worldbank.org/curated/en/099408408092319382/pdf/IDU0301a3e

World Business Council for Sustainable Development. (2023). Leading the Way: CEO Guide to Aligning with the SDGs.

World Economic Forum. (2023). Smart waste management. https://www.weforum.org/

World Food Programme (WFP). (2023). Climate Change and Food Security. https://www.wfp.org/climate-action

World Health Organization. (2023, October 18). Ambient air pollution. https://www.who.int/news-room/fact-sheets/detail/ambient-%28outdoor%29-air-quality-and-health

World Wildlife Fund. (2023, March 8). Sustainable agriculture: How to feed the world without destroying the planet. https://www.worldwildlife.org/blogs/sustainability-works?blog_category_id=food-and-agriculture

Xinhua. (2020, September 22). Xi Jinping announces China's goals to achieve carbon neutrality by 2060.

https://english.news.cn/20221109/465354339ce246bcac743196e213e044/c.html

Zero Hour. (n.d.). Our Story. https://thisiszerohour.org/

Zero Waste International Alliance. (n.d.). Zero Waste International Alliance. https://zwia.org/

www.ingramcontent.com/pod-product-compliance
Lightning Source LLC
LaVergne TN
LVHW022041200926
848481LV00003B/728
* 9 7 9 8 2 2 7 1 4 1 1 7 0 *